T0211580

Communications in Computer and Information Science 765

Commenced Publication in 2007
Founding and Former Series Editors:
Alfredo Cuzzocrea, Xiaoyong Du, Orhun Kara, Ting Liu, Dominik Ślęzak,
and Xiaokang Yang

More information about this series at http://www.springer.com/series/7899

Tibor Bosse · Bert Bredeweg (Eds.)

BNAIC 2016: Artificial Intelligence

28th Benelux Conference on Artificial Intelligence
Amsterdam, The Netherlands, November 10–11, 2016
Revised Selected Papers

 Springer

Editors
Tibor Bosse
Department of Computer Science
Vrije Universiteit Amsterdam
Amsterdam
The Netherlands

Bert Bredeweg
Informatics Institute
University of Amsterdam
Amsterdam
The Netherlands

ISSN 1865-0929   ISSN 1865-0937   (electronic)
Communications in Computer and Information Science
ISBN 978-3-319-67467-4   ISBN 978-3-319-67468-1   (eBook)
DOI 10.1007/978-3-319-67468-1

Library of Congress Control Number: 2017953413

Printed on acid-free paper

This Springer imprint is published by Springer Nature
The registered company is Springer International Publishing AG
The registered company address is: Gewerbestrasse 11, 6330 Cham, Switzerland

Preface

This book contains a selection of the best papers presented at the 28th edition of the annual Benelux Conference on Artificial Intelligence (BNAIC 2016). BNAIC 2016 took place during November 10–11 in Hotel Casa 400 in Amsterdam. The conference was jointly organized by the University of Amsterdam and the Vrije Universiteit Amsterdam, under the auspices of the Benelux Association for Artificial Intelligence (BNVKI) and the Dutch Research School for Information and Knowledge Systems (SIKS).

BNAIC 2016 was the 28th edition of a conference series that started in 1988 under the name "Netherlands Artificial Intelligence Conference." Originally focusing on Artificial Intelligence research in The Netherlands, the conference later expanded its scope to include research in Belgium (since 1999) and Luxembourg (since 2008). Hence, the objective of BNAIC is to promote and disseminate recent research developments in Artificial Intelligence, particularly within Belgium, Luxembourg, and The Netherlands, although it does not exclude contributions from countries outside the Benelux. Part of the success of the BNAIC series can be attributed to the fact that the conference typically receives a large number of excellent student papers, which reflects the high quality of Artificial Intelligence education in the Benelux.

The BNAIC 2016 program was very exciting and diverse: in addition to the regular research presentations, posters and demonstrations (discussed below), it included several other elements, among which were: (a) keynote presentations by Marc Cavazza (University of Kent), Frank van Harmelen (Vrije Universiteit Amsterdam), Hado van Hasselt (Google DeepMind), and Manuela Veloso (Carnegie Mellon University); (b) a Research Meets Business session; (c) a panel discussion on Social Robots, with contributions by Elly Konijn (Vrije Universiteit Amsterdam), Ben Kröse (University of Amsterdam and Amsterdam University of Applied Sciences), Mark Neerincx (TNO and Delft University of Technology), and Peter Novitzky (University of Twente); (d) a special FACt (FACulty focusing on the FACts of AI) session with presentations by Bart de Boer (Vrije Universiteit Brussels), Catholijn Jonker (Delft University of Technology), and Leon van der Torre (University of Luxembourg); and (e) a special session on open access publishing with contributions by Rinke Hoekstra (Vrije Universiteit), Maarten Frohlich (IOS Press), Bernard Aleva (Elsevier), and Hilde van Wijngaarden (University of Amsterdam and Amsterdam University of Applied Sciences).

In addition, BNAIC 2016 featured three awards. The SNN Best Paper Award was bestowed to Rik van Noord, Florian Kunneman, and Antal van den Bosch for their contribution "Predicting Civil Unrest by Categorizing Dutch Twitter Events," which is included in this volume. The SKBS Best Demo Award was given to Caitlin Lagrand, Patrick M. de Kok, Sébastien Negrijn, Michiel van der Meer, and Arnoud Visser for their contribution "Autonomous Robot Soccer Matches." The Best Thesis Abstract

Award was awarded to Hossam Mossalam for his contribution "Multi-Objective Deep Reinforcement Learning with Optimistic Linear Support."

As in previous years, BNAIC 2016 welcomed four types of contributions, namely, (a) regular papers, (b) compressed contributions, (c) demonstration abstracts, and (d) thesis abstracts. The conference received 93 submissions, consisting of 24 regular papers, 47 compressed contributions, 11 demonstration abstracts, and 11 thesis abstracts. After thorough review by the Program Committee, the conference chairs made the final acceptance decisions. The overall acceptance rate for presentation at the conference was 88% (63% for regular papers, 100% for compressed contributions and demonstration abstracts, and 91% for thesis abstracts). Among these accepted contributions, a small subset of the best papers were selected (after another substantial review and revision phase) for inclusion in this volume. The overall acceptance rate was 28% (33% for regular papers, 9% for demonstration abstracts, and 36% for thesis abstracts; compressed contributions were not considered for this volume, as they describe research that has already been published). This is the first year in which a selection of BNAIC papers is published in the form of a separate Springer book, and our ambition is to continue this initiative in the coming years.

The contributions presented in this volume are grouped by category and by topic. First, eight regular papers are included, of which the first four are all related to natural language processing. The next regular paper presents an agent-based approach to image reconstruction. After that, two regular papers are included that address aspects of game theory, and the last regular paper proposes a solution to the travelling umpire problem based on answer set programming. The regular papers are followed by four student papers, of which the first presents a fuzzy logic approach for anomaly detection in energy consumption data. The next two student papers describe interesting studies regarding the interaction between artificial agents and humans, and the last student paper reports on a learning analytics case study in the context of online secondary education. The last paper included in the book originates from a demonstration abstract and describes an integrated collaborative environment for data processing, exploration and analysis.

To conclude, we want to express our gratitude to everyone who contributed to this book and to BNAIC 2016: in addition to all the invited speakers mentioned above, many thanks to all Organizing and Program Committee members for their hard work in assuring the high quality of the conference and the proceedings. Moreover, we wish to thank all student volunteers, administrative and secretarial assistants, and of course our sponsors. We also gratefully acknowledge help from the BNVKI and from previous BNAIC organizers. And last, but certainly not least, we cordially thank all the authors who submitted their important contributions. Without their efforts, this volume could not have been published.

July 2017

Tibor Bosse
Bert Bredeweg

Organization

General Chairs

Tibor Bosse	Vrije Universiteit Amsterdam, The Netherlands
Bert Bredeweg	University of Amsterdam, The Netherlands

Student Program

Arnoud Visser	University of Amsterdam, The Netherlands

Interactive Demos

Tom Kenter	University of Amsterdam, The Netherlands

Contact and Administration

Mojca Lovrenčak	Vrije Universiteit Amsterdam, The Netherlands

Sponsoring

Natalie van der Wal	Vrije Universiteit Amsterdam, The Netherlands

Finances

Mark Hoogendoorn	Vrije Universiteit Amsterdam, The Netherlands

Website

Adnan Manzoor	Vrije Universiteit Amsterdam, The Netherlands

Program Committee

Stylianos Asteriadis	University of Maastricht, The Netherlands
Reyhan Aydogan	Delft University of Technology, The Netherlands
Floris Bex	Utrecht University, The Netherlands
Michael Biehl	University of Groningen, The Netherlands
Mauro Birattari	IRIDIA, Université Libre de Bruxelles, Belgium
Peter Bloem	University of Amsterdam, The Netherlands
Sander Bohte	Centrum Wiskunde & Informatica, The Netherlands
Peter Bosman	Centrum Wiskunde & Informatica, The Netherlands
Tibor Bosse	Vrije Universiteit Amsterdam, The Netherlands
Bruno Bouzy	Paris Descartes University, France

Makiko Sadakata	University of Amsterdam, The Netherlands
Stefan Schlobach	Vrije Universiteit Amsterdam, The Netherlands
Pierre-Yves Schobbens	University of Namur, Belgium
Johannes Scholtes	University of Maastricht, The Netherlands
Martijn Schut	AMC, University of Amsterdam, The Netherlands
Evgueni Smirnov	Maastricht University, The Netherlands
Matthijs T.J. Spaan	Delft University of Technology, The Netherlands
Jennifer Spenader	University of Groningen, The Netherlands
Ida Sprinkhuizen-Kuyper	Radboud University Nijmegen, The Netherlands
Pieter H.M. Spronck	Tilburg University, The Netherlands
Thomas Stützle	Université Libre de Bruxelles, Belgium
Johan Suykens	University of Leuven, Belgium
Niels Taatgen	University of Groningen, The Netherlands
Annette ten Teije	Vrije Universiteit Amsterdam, The Netherlands
Dirk Thierens	Universiteit Utrecht, The Netherlands
Jan Treur	Vrije Universiteit Amsterdam, The Netherlands
Karl Tuyls	University of Liverpool, UK
Antal van Den Bosch	Radboud University Nijmegen, The Netherlands
Egon L. van den Broek	Utrecht University, The Netherlands
Jaap van Den Herik	Leiden University, The Netherlands
Wil van der Aalst	Eindhoven University of Technology, The Netherlands
Peter van der Putten	LIACS, Leiden University and Pegasystems, The Netherlands
Leon van der Torre	University of Luxembourg, Luxembourg
Natalie Van Der Wal	Vrije Universiteit Amsterdam, The Netherlands
Tom van Engers	University of Amsterdam, The Netherlands
Tim van Erven	Leiden University, The Netherlands
Frank Van Harmelen	Vrije Universiteit Amsterdam, The Netherlands
Sietse van Netten	University of Groningen, The Netherlands
Martijn Van Otterlo	(former) Radboud University Nijmegen, The Netherlands
M. Birna van Riemsdijk	Delft University of Technology, The Netherlands
Peter van Rosmalen	Open University of The Netherlands, The Netherlands
Maarten Van Someren	University of Amsterdam, The Netherlands
Marieke van Vugt	University of Groningen, The Netherlands
Joost Vennekens	University of Leuven, Belgium
Katja Verbeeck	Odisee, Belgium
Bart Verheij	University of Groningen, The Netherlands
Arnoud Visser	University of Amsterdam, The Netherlands
Louis Vuurpijl	Radboud University Nijmegen, The Netherlands
Willem Waegeman	Ghent University, Belgium
Martijn Warnier	Delft University of Technology, The Netherlands
Gerhard Weiss	University Maastricht, The Netherlands
Max Welling	University of Amsterdam, The Netherlands
Marco Wiering	University of Groningen, The Netherlands
Floris Wiesman	AMC, University of Amsterdam, The Netherlands
Jef Wijsen	University of Mons, Belgium

Mark H.M. Winands	Maastricht University, The Netherlands
Radboud Winkels	University of Amsterdam, The Netherlands
Cees Witteveen	Delft University of Technology, The Netherlands
Yingqian Zhang	Eindhoven University of Technology, The Netherlands

Additional Reviewers

Franz, Robin	Wolf, Ben
Merhej, Elie	Ye, Qing Chuan
Voulis, Nina	

Contents

Demonstration Papers

Regular Papers

Predicting Civil Unrest by Categorizing Dutch Twitter Events

Rik van Noord[1]([⊠]), Florian A. Kunneman[2], and Antal van den Bosch[2,3]

[1] CLCG, University of Groningen, Groningen, The Netherlands
r.i.k.van.noord@rug.nl
[2] Centre for Language Studies, Radboud University, Nijmegen, The Netherlands
{f.kunneman,a.vandenbosch}@let.ru.nl
[3] Meertens Institute, Amsterdam, The Netherlands

Abstract. We propose a system that assigns topical labels to automatically detected events in the Twitter stream. The automatic detection and labeling of events in social media streams is challenging due to the large number and variety of messages that are posted. The early detection of future social events, specifically those associated with civil unrest, has a wide applicability in areas such as security, e-governance, and journalism. We used machine learning algorithms and encoded the social media data using a wide range of features. Experiments show a high-precision (but low-recall) performance in the first step. We designed a second step that exploits classification probabilities, boosting the recall of our category of interest, social action events.

Keywords: Civil unrest · Event categorization · Event detection

1 Introduction

Many instabilities across the world develop into civil unrest. Unrest often materializes into crowd actions such as mass demonstrations and protests. A prime example of a mass crowd action in the Netherlands was the *Project X* party in Haren, Groningen, on September 21, 2012. A public Facebook invitation to a birthday party of a 16-year old girl ultimately led to thousands of people rioting [18]. The riots could only be stopped by severe police intervention, resulting in more than 30 injuries and up to 80 arrests. Afterwards it was concluded that the police were insufficiently prepared and that they were not well enough informed about the developments on social media. An evaluation committee recommended the development of a nation-wide system able to analyze and detect these threats in advance [13]. In this paper, we describe a system that leverages posts on Twitter to automatically predict such civil unrest events before they happen.

To facilitate this objective we start from a large set of open-domain events that were automatically detected from Twitter from a period spanning multiple

T. Bosse and B. Bredeweg (Eds.): BNAIC 2016, CCIS 765, pp. 3–16, 2017.
DOI: 10.1007/978-3-319-67468-1_1

years by the approach described in [9][1]. From this set we aim to identify the events that are materializations of civil unrest, henceforth *social action events*. Arguably, a system that detects social actions should not only reliably detect events where large groups of people come together, but should also exclude events where people gather for different reasons (e.g. soccer matches, music performances) and for which authorities are sufficiently prepared. In addition, a system able to detect several categories reliably might be useful for other applications as well, such as presenting tourists with events of a certain type and in a certain time range. For these reasons, instead of focusing on this event type only, we categorize all events into a broad categorization of events, and distinguish social actions as one of the event types.

This paper is structured as follows. In Sect. 2 we provide an overview of related work, discussing both the tasks of predicting social action events and categorizing Twitter events. In Sect. 3 we present the experimental set-up, discussing the data, event annotations and event classification. We present the results of our system evaluation in Sect. 4, and analyze the retrieval of social action events, as well as the most informative features, in Sect. 5. Conclusions and a discussion are presented in Sect. 6.

2   Related Work

2.1   Predicting Social Action Events

There is a small body of work dedicated to detecting social action events. [3] aim to predict civil unrest in South America based on Twitter messages. In contrast to our approach, they predict such events directly from tweets, by matching them with specific civil unrest related keywords, a date mention, and one of the predefined locations of interest. Their system obtains a precision of 0.55 on a set of 283 predefined events. The main drawback of their approach is that it has no predictive abilities. For example, the system is not able to detect social action events that use newly emerging keywords for a specific event, or take place in a new location. As a consequence, their system has a low recall; many future social actions are likely to go undetected.

A more generic approach to detecting social action events is the EMBERS system by [16]. They try to forecast civil unrest by using a number of open source data sources such as Facebook, Twitter, blogs, news media, economic indicators, and even counts of requests to the TOR browser.[2] Using multiple models, the system issues a warning alert when it believes a social action event is imminent. Tested over a month, the EMBERS system attained a precision of 0.69 and a recall of 0.82. [6] provide a more detailed explanation of some of the EMBERS models, reporting similar F-scores as [16]. They also compare

---

[1] A live event detection system using the method of [9] is available at http://lamaevents.cls.ru.nl/.

[2] TOR requests are an indication of the number of people who choose to hide their identity and location.

the impact of different data sources, concluding that Social Media (including Twitter, blogs and news) is the most informative source. [12] test EMBERS when only taking Twitter information into account, reporting a precision of 0.97 but a recall of 0.15.

2.2 Categorizing Events

Some approaches based on Twitter perform some form of broad categorization of events [15,20]. These approaches either identify which topics are often talked about on Twitter, or focus on the categorization of users instead of events. To our knowledge, the only approach that focuses on the categorization of automatically detected events is the one proposed by [17]. They apply Latent Dirichlet Allocation [2] to a set of 65 million events to generate 100 topical labels automatically. Manual post-annotation winnowed these down to a set of 37 meaningful categories. 46.5% of the events belong to one of these categories, while 53.5% of the events are in a rest category. [17] compared their unsupervised approach to categorizing Twitter events to a supervised approach. They selected the best 500 events (detected with the highest confidence) and manually annotated them by event type. Their unsupervised approach obtained an F1-score of 0.67, outperforming the supervised approach which obtained an F1-score of 0.59. However, they do show that the F1-score of the supervised approach steadily increases when using more training instances.

3 Experimental Set-Up

Our study starts with a set of automatically detected events from Twitter, described in Sect. 3.1. We manually annotate two subsets of these events by type, and subsequently train a machine learning classifier on several feature types extracted from these events. Performance is both evaluated on the annotated event sets and on the larger set of remaining events.

3.1 Data

Event Set. To perform automatic event categorization, we use the event set described in [8] which was extracted based on the approach described in [9]. As this approach was applied to Dutch tweets, the set mainly comprises Dutch events. This approach, based on the method of [17], comprises the extraction of explicit time expressions and entities from tweets, identifying date-entity pairs as event when they co-occur together for at least five times and display a good fit as measured by the $G2$ log likelihood ratio statistic.

An example of a detected social action event on Twitter is shown in Table 1 [19]. Each event has a set of attributes, such as the date, keywords, tweets and event score. The event score is linked to the size and popularity of an event. For the exact calculation of this score, we refer to [9, pp. 13]. Over a 6-year period (2010–2015), [8] ultimately obtained 93,901 events. This event set is used for our categorization system.

Table 1. An example of an actual Dutch social action event with five example tweets (translated to English).

Date	21-09-2013
Keywords	#demonstration, budget cuts
21 September: say no to the new budget cuts! #demonstration	
Are you also coming to the #demonstration #21september ? #action is necessary!	
Come Sept 21 to The Hague to demonstrate against the cabinet #demonstration	
It is allowed again tomorrow, no excuses not to go to #thehague #resistance	
8 days to go #PVV #demonstration against #cabinet at #koeplein The Hague!	

Event Annotations. We select two sets of events for manual labeling. Our first event set contains the 600 events with the highest event score in the output of [8]. This enables us to make an approximate comparison to [17], who evaluated their system on the basis of their 500 top-ranked events. We refer to the set of events with the highest event scores as the *best event set*.

Our second event set is created by randomly selecting an event from the ranked total event set for intervals of 155 events (with all events ranked by event score), excluding the best 600 events of the best event set. We refer to this event set as the *random event set*. Non-Dutch events were manually removed from both event sets, leaving 586 events in the best events set and 585 in the random events set.

First, we annotated the set of best events. Seven annotators were involved in the annotation process, who all at least annotated 40 and at most 175 of the events in the best event set. 195 of the 586 best events received a double annotation so that we are able to calculate inter-annotator agreement. The other 390 events, as well as the 585 random events, were annotated by one annotator. Similar to [17], the annotator is asked two questions for each event:

- Is this an actual event according to the definition?
- What is the category of this event?

We employ the same definition as [9] as to what constitutes an actual event: 'An event is a significant thing that happens at some specific time and place', where 'significant' is defined as 'something that may be discussed in the news media'. An event in our full event set is not necessarily a proper event according to this definition, as the detection procedure makes errors. Since we are not interested in the category of a non-event, the events that are annotated as a non-event are filtered from the event set.

We defined ten possible categories after an initial manual inspection of about 200 events. They are listed in Table 2. *Social action* is the category of interest.

Table 2. The ten different categories with examples.

Category	Example events
Social action	Strikes, demonstrations, flashmobs
Sport	Soccer match, local gymnastics event
Politics	Election, public debate
Broadcast	Television show, premiere of a movie
Public event	Performance of a band, festival
Software	Release of game, release of new iPhone
Special day	Mother's Day, Christmas
Celebrity news	Wedding or divorce of a celebrity
Advertisement	Special offers, retweet and win actions
Other	Rest category

As arguably less straightforward categories we included *special day* and *advertisement* because manual inspection of the data suggested that those types of events were frequent enough to deserve their own category.

The 195 events annotated by two coders yielded a Krippendorff's alpha [5] of 0.81 on judging whether or not it was an actual event and 0.90 on categorizing events. These scores can be considered excellent [7] and show that we can reliably view the events that were annotated once as if they were annotated correctly. Therefore, the 586 random events could be annotated once by two annotators.

Events that were (at least once) annotated as a non-event are removed from the event set, as well as events where annotators disagreed on the category. 27.4% of the best events were a non-event, leaving 425 of the best events. In the random event set 38.1% were discarded as non-events, leaving 362 events.

The annotations by event category are shown in Fig. 1. *Public event* is the dominant category, comprising 29.6% of the best events and 44.5% of the random events. Most other event categories occur fairly regularly, except *advertisement* and *celebrity news*. The latter category was so infrequent that it was removed from both event sets. *Advertisement* was removed from the random event set, but was retained for the best event set.

3.2 Training and Testing

Based on the annotated events we trained a machine learning classifier to distinguish the ten event types. We describe the event features, classification approaches and evaluation below.

Feature Extraction. To enable the classifier to learn the specific properties of each event category we extract several types of features from each event. They are listed in Table 3. The first four feature types are derived from [9]. The first feature type, the event score, describes the link between the event keywords and

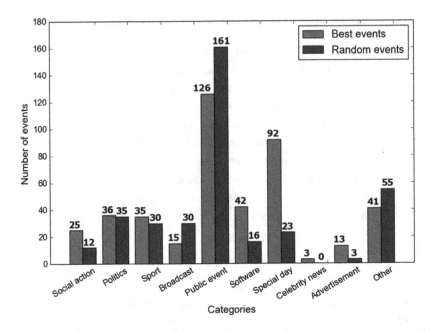

Fig. 1. The ten different categories with the number of annotated examples in the best and random event set.

the date of the event. This score gives an indication of the confidence that the set actually represents an event. Second, the keyword scores give an indication of the commonness of each event keyword, based on the commonness score as described in [10]. Third, the event date might help to recognize event types that are linked to big events, such as elections. Fourth, we extract the number of tweets, which might reflect the popularity of an event. We also distinguish between the numbers of tweets before, during, and after[3] an event. Fifth, we extract each word used in the event tweets as a feature, jointly referred to as bag-of-words features. Such features provide the classifier with a lot of information, but there is no deeper reasoning involved concerning the words in question. The most informative words per category are shown in Table 4. As a sixth feature type we scored the average subjectivity and polarity of each event tweet, using the approach by [4]. The subjectivity and polarity score of the event are averaged over the scores of all event tweets. Some event types might be referred to fairly objectively in tweets, while others might stir more sentiment.

---

[3] Since we wanted to provide our system with as much training data as possible, we also extracted relevant tweets that were posted after the event took place. Obviously, when predicting events in the future, this type of data will be unavailable.

Table 3. Types of extracted features with descriptions.

Feature type	Description
Event score	Single feature specifying the event-score
Keyword scores	Feature per keyword specifying the keyword-score
Event date	Single feature specifying the event date
Tweet count	Three features, specifying the total number of tweets and number of tweets before and after the event
Bag-of-words	Each unique word has its own feature, the value of each feature is determined by how often the word occurs in the tweets of the event
Sentiment	Two features: the average subjectivity and the average polarity, calculated over all tweets of the event
Periodicity	Two features: one binary feature that specifies if the event is periodic and one feature that specifies the periodicity type (e.g. yearly)
Wikipedia	Each unique Wikipedia type has its own feature, the value of each feature is determined by how often the type occurs in the tweets of the event

The seventh feature type indicates whether an event is of a periodic nature. This feature is based on the output from a periodicity detection system described by [8]. Finally, we employ DBpedia [1] in order to generalize over the different named entities present in the events. Since we want to generalize over the different terms, we are especially interested in the **type** attribute of the entity in DBpedia. This gives us a broader description of the named entities in question and therefore allows for generalization of previously unseen entities. For example, Feyenoord is a *SoccerClub*, *SportsTeam* and *Organisation*, while Justin Bieber is an *Artist*, *MusicalArtist*, *MusicGroup*, and a *NaturalPerson*. We extract the different DBpedia **types** for each event keyword. The keywords are linked to DBpedia using Wikification [11].

Table 4. An ordered list of the 8 most indicative words per category according to their *tf * idf* score (translated from Dutch to English).

Category	Most indicative words
Social action	Against everyone protest respect they demonstration all
Politics	Votes elections vote cda vvd pvda d66 groenlinks
Sport	Match against soccer wins rt ajax psv tonight
Broadcast	Tv watch tonight episode see tvtip show season
Public event	rt tonight was today what who much tomorrow
Software	Apple iphone microsoft out gta wait comes windows
Special day	Today rt what everyone day on if celebrate
Other	Not rt that no what will so today one

Classification. Based on the extracted feature sets along with the annotated categories, we train a Naive Bayes classifier[4] using the Python module Scikit-learn[5] [14]. We use Laplace smoothing ($\alpha = 1.0$) and learn the prior probabilities per class. No correction method for document length is employed.

We applied two methods to increase the performance of the classifier: down-sampling the dominant *public event* class, and performing bag-of-words classification as a first-step classification. The first method simply reduces the number of *public events* in the training set to ensure it does not hinder the performance of the minority classes. The number of *public events* is reduced to the same frequency as the second-most frequent class in the event set, resulting in the deletion of 34 events in the best event set and 106 events in the random event set.

The second method only feeds the bag-of-words features to the classifier in an initial stage, and subsequently adds the resulting classification to the set of other features. The advantage of this stacking method is that it reduces the dominance of the word features compared to the other features, allowing the classifier to view the word features as a single source of information. Also, it enables us to measure the impact of the non bag-of-words features in comparison to a bag-of-words baseline.

Evaluation. The performance on categorizing events is evaluated in two ways. The first is to apply 5-fold cross validation on the annotated sets of events. We do this for both the best event set (for a comparison with [17]) and the random event set, calculating the average precision, recall and F1-score.[6] The second way is to evaluate the results on a set that was never used in the training phase. The classifiers are trained on the two sets of annotated events and subsequently applied to the remaining 92,701 events. Performance on these unseen events is evaluated by manually inspecting a subset of them. As *public event* appeared to be a very dominant category, occurring 81,538 times in the full set of events according to the classifier, it was not feasible to randomly select a set of events to be used as evaluation set. This is why we focus on evaluating the precision of each classification category separately. We randomly selected 50 events per category for evaluation, except for our category of interest, social action events, for which we include all 93 events classified with this category. *Advertisement* could only be evaluated for 25 events, as it was only predicted 25 times. This ultimately resulted in a total set of 468 events, which we refer to as the *Evaluation set*.

---

[4] In addition to Naive Bayes, we experimented with Support Vector Machines and K-nearest neighbors. We will only report on the outcomes of Naive Bayes, which yielded the best performance.

[5] http://scikit-learn.org.

[6] This was calculated by using the *weighted* setting in scikit-learn, which is why the F-score is not necessarily between precision and recall.

4   Results

4.1   Annotated Set

Table 5 shows the most important results of the 5-fold cross validation. Averaged over all categories, the best event set obtained an F1-score of 0.65, while the random event set received an F1-score of 0.58. It appears to be easier to classify events with a higher event score. However, we found no significant effect of event score when doing a least-squares logistic regression test for the random event set ($r(360) = -0.05$, $p = 0.39$). This suggests that there is a small subset of events with a very high event score that is easier to classify, but that there is no significant effect of event score in general.

Comparing the setting where only bag-of-words is used as a feature with the setting where the classification based on bag-of-words is added as a feature to the other features, the latter setting yields the best outcomes.

The score on our best event set is similar to the score of [17]. However, it is hard to make a fair comparison, since they did not include a category distribution of the test set in their 37-class problem.[7]

Social action is predicted at a high precision in the best events set, but the scores for the random events are poor. This might be due to the low number of instances in this set (12), in comparison with the 25 social actions in the best event set.

Table 5. The results of the 5-fold cross validation for the Naive Bayes algorithm while down-sampling the dominant *public event* class.

		All categories			Social actions		
		Prec	Rec	F1	Prec	Rec	F1
Best events	Only bag-of-words	0.67	0.59	0.55	0.68	0.41	0.52
	Bag-of-words as feature	**0.67**	**0.67**	**0.65**	0.79	0.44	0.56
Random events	Only bag-of-words	0.61	0.60	0.57	0.40	0.17	0.24
	Bag-of-words as feature	**0.64**	**0.60**	**0.58**	0.40	0.17	0.24

Down-sampling increased the F1-score by 0.05 for the best event set and 0.06 for the random event set.

4.2   Evaluation Set

Table 6 shows the results on the Evaluation set, listing the precision per category. In general, these scores are high for a 9-class classification task. The precision per class is even 1.00 for *sport* and *politics*, meaning that if the classifier predicted those categories, it did so perfectly. The categories *public event* and *advertisement* score below 0.70, however. The low precision for *public event* impacts the

---

[7] In personal communication, we asked Alan Ritter about this distribution. Unfortunately, he was unable to recover the document with the specific division of categories in the test set.

Table 6. The precision and number of predicted instances per category.

Category	Instances	Precision	Category	Instances	Precision
Social action	93	0.80	Software	1,630	0.96
Politics	2,170	0.86	Special day	1,722	0.78
Sport	2,771	1.00	Advertisement	25	0.51
Broadcast	206	1.00	Other	1,535	0.70
Public event	81,538	0.57			

overall performance of the classification system substantially. As 81,538 out of 92,701 events were classified as a *public event*, a precision of 0.57 leads to about 35 thousand incorrectly classified events.

We should keep in mind that the non-events were not excluded from the full event set. It was estimated that 38.1% of all detected events are not events. In the training phase these non-events were excluded, so it is likely that the classifier will assign many non-events in the full event set to the most frequent category. A large part of the bias to *public event* may be due to the occurrence of non-events in the full event set. This leads us to conclude that if there were a more reliable way to automatically exclude non-events, the results of the general categorization would considerably improve.

The results for the *Social Action* category are promising, since the 93 *social actions* in this set were predicted with a precision of 0.80. However, we estimate that the recall of this category will be low. Only 93 out of 92,701 events (0.1%) were predicted as a *social action*, while 3.3% of events were annotated as a *social action* in the random event set.

5   Analysis

5.1   Increasing the Recall for Social Action Events

Our main goal is to detect *social action events* and possibly alerting the authorities when such an event will take place. Therefore, we rather show a large list of events that might be a social action event that actually includes most of the actual events, than a system that often misses them. Since we are not talking about thousands of events daily, an analyst could annotate the set of possible social action events manually. We thus prefer a high recall to a high precision. Therefore, we propose a method to increase the recall of social action events, at minimal precision costs.

In order to increase recall we make use of the Naive Bayes classifier probability by category that is assigned to each event. Events for which *social action* obtained the second highest probability are ignored by default, as another category is picked. One way to remedy this is to classify all events where *social action* was the second most probable class. We refer to these events as **secondary social action events**. By doing this we were able to expand this set with 226 additional events, which we annotated manually. 26 of the 226 *secondary social actions* were annotated as a non-event and were thus excluded from the set. 130 of the remaining 200 events were indeed annotated as a social action, resulting in a precision of **0.65**. Adding the 200 events to the *social action* events in the evaluation set results in a drop of total precision from 0.80 to 0.69. Thus, since we could add 130 *social action events* to the 56 that were already found as a primary classification, the recall was increased by **232%** while the precision only dropped by **14%**. Hence, including the *secondary social action events* seems a useful method for increasing the recall, while only mildly hurting precision.

A possible other method that exploits the actual Bayesian probabilities would be to select all events for which the probability of *social action event* exceeds a certain threshold. This would then allow us to pick a specific precision-recall trade-off, instead of simply relying on events that were second in the ranking of probabilities. Investigating this is left for future work.

5.2   Most Informative Features

In order to achieve some insight from the most informative features for the two event sets, we calculated the chi-squared value for each feature in relation to the category label. These are listed in Table 7. The most informative features are generally intuitive. They include words such as *stemmen (to vote)* and *stem (vote)* as indicators of a political event, but also specific hashtags such as *#VVD* and *#CDA*; CDA and VVD are political parties in The Netherlands. The best predictors for *sport* are the DBpedia type features *SoccerClub* and *ClubOrganization*. The most indicative features of the category *social action* are the words *protest* and *demonstratie (demonstration)*. Although these words almost exclusively occurred in *social action events*, due to their low frequency they do not rank in the feature top 100.

Table 7. The eight best features for the best and random event set, based on their chi-squared value. Non-word features are in italics. Features are only included if they occurred at least ten times in their event set.

Best events		Random events	
Feature	Category	Feature	Category
stemmen (vote)	Politics	*ClubOrganization*	Sport
stem (vote)	Politics	*SoccerClub*	Sport
19-03-2014	Politics	wint (wins)	Sport
SoccerClub	Sport	wedstrijd (match)	Sport
#vvd	Politics	2015	Politics
wedstrijd (match)	Sport	seizoen (season)	Broadcast
ClubOrganization	Sport	tv	Broadcast
#cda	Politics	tegen (against)	Sport

The polarity, subjectivity and periodicity features turned out to be less valuable, ranking in the bottom 25% of all features. This is surprising, since *special days* are often periodic, while it is, for example, uncommon for *social action events* to be periodic.

6   Conclusion and Discussion

In this study we presented a generic event categorization system which we evaluated particularly on its ability to predict civil unrest. The general categorization system has a bias towards the dominant category *public event*, but has a high precision for the other categories, including *social action*. The recall for *social action* was low; a follow-up step that exploited the specific per-class probabilities generated by the Naive Bayes classifier led to a considerable improvement in recall of 232%, at the minor cost of a 14% decrease in precision.

The study by [17] is the only related study that also produced an extensive evaluation of event categorization, evaluating their system on a set of 500 events with the highest association (similar to the event score by which we selected a set of best events). Their 37-class approach ultimately obtained a precision, recall and F1-score of 0.85, 0.55 and 0.67. Our system offered a comparable performance: a precision, recall and F1-score of 0.67, 0.67 and 0.65.

A comparable approach to predicting civil unrest is the EMBERS system by [16]. They evaluated their system over a period of a month, resulting in a precision and recall of respectively 0.69 and 0.82. In comparison, we obtained a higher precision while our estimated recall is lower. It is interesting to note how they received this recall score. They obtained a gold standard set of *social action events* by an independent organization that had human analysts survey newspapers and other media for mentions of civil unrest; arguably a reliable way of calculating recall in the real world. Our approach is only able to recall events that were present in the set of [8]. We have not explored ways to evaluate to what extent [8] detected all *social action events* that actually happened. We should consider the possibility that we might still miss *social action events* that were never detected as events in the first place, lowering our estimated recall.

Using the ranking of the Bayesian probabilities helped to increase the recall of *social action events* by 232%. We did not use the actual probabilities to influence the classification process, but used only the ranking of these probabilities. A potential direction for future research is to use the per-class probabilities generated by the Naive Bayes classifier in a more sophisticated manner. For example, it is possible to learn a certain probability threshold for *social action* and classify events that exceed this threshold as *social action*, regardless of the probability of other categories. The actual implementation of such a method requires a search for the best threshold setting. The main advantage of this approach is that this allows us to specify a specific precision-recall trade-off that is the most suitable for predicting social action events.

Our study has shown that the detection of *social action events* from Twitter based on open-domain event extraction and a subsequent event categorization procedure is feasible. Due to the broad scope on open-domain events as starting point, we expect that this approach could be refined and improved when the focus is more on social action, e.g. by using lexicons of words associated with social action. Studying the extent of the potential added value of domain-specific knowledge is open for future work.

References

1. Bizer, C., Lehmann, J., Kobilarov, G., Auer, S., Becker, C., Cyganiak, R., Hellmann, S.: Dbpedia-a crystallization point for the web of data. Web Semant. Sci. Serv. Agents World Wide Web **7**(3), 154–165 (2009)
2. Blei, D.M., Ng, A.Y., Jordan, M.I.: Latent Dirichlet allocation. J. Mach. Learn. Res. **3**, 993–1022 (2003)
3. Compton, R., Lee, C.-K., Lu, T.-C., de Silva, L., Macy, M.: Detecting future social unrest in unprocessed Twitter data: emerging phenomena and big data. In: 2013 IEEE International Conference on Intelligence and Security Informatics (ISI), pp. 56–60. IEEE (2013)
4. De Smedt, T., Daelemans, W.: Pattern for Python. J. Mach. Learn. Res. **13**(1), 2063–2067 (2012)
5. Hayes, A.F., Krippendorff, K.: Answering the call for a standard reliability measure for coding data. Commun. Methods Meas. **1**(1), 77–89 (2007)
6. Korkmaz, G., Cadena, J., Kuhlman, C.J., Marathe, A., Vullikanti, A., Ramakrishnan, N.: Multi-source models for civil unrest forecasting. Soc. Netw. Anal. Min. **6**(1), 1–25 (2016)
7. Krippendorff, K.: Content Analysis: An Introduction to Its Methodology. Sage, Thousand Oaks (2004)
8. Kunneman, F., van den Bosch, A.: Automatically identifying periodic social events from Twitter. In: Proceedings of the RANLP 2015, pp. 320–328 (2015)
9. Kunneman, F., van den Bosch, A.: Open-domain extraction of future events from Twitter. Nat. Lang. Eng. **22**, 655–686 (2016)
10. Meij, E., Weerkamp, W., de Rijke, M.: Adding semantics to microblog posts. In: Proceedings of the Fifth ACM International Conference on Web Search and Data Mining, pp. 563–572. ACM (2012)
11. Mihalcea, R., Csomai, A.: Wikify!: linking documents to encyclopedic knowledge. In: Proceedings of the Sixteenth ACM Conference on Information and Knowledge Management, pp. 233–242. ACM (2007)
12. Muthiah, S., Huang, B., Arredondo, J., Mares, D., Getoor, L., Katz, G., Ramakrishnan, N.: Planned protest modeling in news and social media. In: AAAI, pp. 3920–3927 (2015)
13. NOS: Cohen: fouten politie, burgemeester. Nederlandse Omroep Stichting, 7 March 2013. http://nos.nl/
14. Pedregosa, F., Varoquaux, G., Gramfort, A., Michel, V., Thirion, B., Grisel, O., Blondel, M., Prettenhofer, P., Weiss, R., Dubourg, V., Vanderplas, J., Passos, A., Cournapeau, D., Brucher, M., Perrot, M., Duchesnay, E.: Scikit-learn: machine learning in Python. J. Mach. Learn. Res. **12**, 2825–2830 (2011)
15. Ramage, D., Dumais, S.T., Liebling, D.J.: Characterizing microblogs with topic models. ICWSM **10**, 1 (2010)

16. Ramakrishnan, N., Butler, P., Muthiah, S., Self, N., Khandpur, R., Saraf, P., Wang, W., Cadena, J., Vullikanti, A., Korkmaz, G., et al.: 'Beating the news' with embers: forecasting civil unrest using open source indicators. In: Proceedings of the 20th ACM SIGKDD International Conference on Knowledge Discovery and Data Mining, pp. 1799–1808. ACM (2014)
17. Ritter, A., Etzioni, O., Clark, S., et al.: Open domain event extraction from Twitter. In: Proceedings of the 18th ACM SIGKDD International Conference on Knowledge Discovery and Data Mining, pp. 1104–1112. ACM (2012)
18. van Heerden, D.: Facebook birthday invite leads to mayhem in Dutch town, authorities say. CNN, 24 September 2012. http://edition.cnn.com/
19. Volkskrant: Enkele duizenden bij protestmars bezuinigingen. de Volkskrant, 21 September 2013. http://volkskrant.nl/
20. Zhao, W.X., Jiang, J., Weng, J., He, J., Lim, E.-P., Yan, H., Li, X.: Comparing Twitter and traditional media using topic models. In: Clough, P., Foley, C., Gurrin, C., Jones, G.J.F., Kraaij, W., Lee, H., Mudoch, V. (eds.) ECIR 2011. LNCS, vol. 6611, pp. 338–349. Springer, Heidelberg (2011). doi:10.1007/978-3-642-20161-5_34

Textual Inference with Tree-Structured LSTM

Adebayo Kolawole John[✉], Luigi Di Caro, Livio Robaldo, and Guido Boella

Dipartimento di Informatica, Universita Di Torino,
Corso Svizzera 185, 10149 Torino, Italy
`kolawolejohn.adebayo@unibo.it`, {`dicaro,guido.boella`}`@di.unito.it`,
`livio.robaldo@uni.lu`

Abstract. Textual Inference is a research trend in Natural Language Processing (NLP) that has recently received a lot of attention by the scientific community. Textual Entailment (TE) is a specific task in Textual Inference that aims at determining whether a hypothesis is entailed by a text. This paper employs the *Child-Sum Tree*-LSTM for solving the challenging problem of textual entailment. Our approach is simple and able to generalize well without excessive parameter optimization. Evaluation done on SNLI, SICK and other TE datasets shows the competitiveness of our approach.

Keywords: Child-Sum Tree LSTM · Information retrieval · Textual entailment

1 Introduction

Natural Language Inference (NLI) or put in another way, Textual Inference, refers to the process of identifying the type of logical/semantic relationship that exists between two texts. Since Dagan et al. [14] conceived the task of Recognizing Textual Entailment (RTE), it has continued to receive a lot of interest from researchers. To be specific, *entailment, contradiction* and *neutral* are examples of the inference relationships that are to be determined. Other examples of language inference tasks include text similarity [17], answer sentence selection [15], as well as Paraphrase Detection [28]. These tasks are challenging due to the prevalent variability and ambiguity in natural languages [13].

The preceeding work in NLI employs typical machine learning (ML) classifiers, this requires a lot of efforts for handcrafting feature for the classifiers. Moreover, these systems also rely on many language resources and tools, e.g., Semantic Nets etc. Also, the tiny size of the datasets involved, i.e., the SICK[1] and RTE[2] corpuses, which were the earliest datasets for RTE evaluation, contributes to the choice of ML classifiers because they have small training samples. This discourages the use of neural networks, which are more data-intensive. Most importantly, these systems rarely or slightly outperform simple baselines which rely on simple surface string similarity and word-overlap approaches [1,4].

---

[1] http://clic.cimec.unitn.it/composes/sick.html.

[2] https://www.aclweb.org/aclwiki/index.php?title=Textual_Entailment_Resource_Pool.

© Springer International Publishing AG 2017
T. Bosse and B. Bredeweg (Eds.): BNAIC 2016, CCIS 765, pp. 17–31, 2017.
DOI: 10.1007/978-3-319-67468-1_2

Premise	Hypothesis	Inference Relationship
This church choir sings to the masses as they sing joyous songs from the book at a church.	A choir singing at a baseball game.	Contradiction
This church choir sings to the masses as they sing joyous songs from the book at a church.	The church has cracks in the ceiling.	Neutral
This church choir sings to the masses as they sing joyous songs from the book at a church.	The church is filled with song.	Entailment

Fig. 1. Sample texts from SNLI showing different classes of inference relationship

NLI is purely a classification task between different classes of inference relationship. Figure 1 shows one example for each of the three classes 'contradiction', 'neutral' and 'entailment' from the SNLI corpus[3].

Since the release of SNLI by Bowman et al. [9], a lot of research work that is based on neural networks have been published. Quite a good number of these systems are based on sentence encoding [9,11,26], where the Long Short-Term Memory (LSTM) networks have been used to embed premises and hypothesis in the same vector space. This enhances parameter sharing throughout the other components of a neural network model. Other techniques like the attention mechanism [20,21,23], extended memory structure [12,22] and factorization-based matching [32] builds on the former by providing more interaction between the embedded sentences.

Even though these systems have reported impressive result, they often exhibit a deep sentence modeling, which often translates to having excessive trainable parameters. Furthermore, the kind of interaction that the systems focus on downplays the syntactic relationship and interplay between the words in each sentence. However, we know that words do not live in isolation, and the meaning of a word is context dependent. Therefore, in order to know the true meaning of a word, it is also required that we know the meaning of the neighouring words that modifies its meaning.

This work employs a Tree-LSTM to obtain the representation of both the Premise[4] and its Hypothesis. Tree-LSTM is a linguistically intuitive choice for it readily captures the syntactic interpretations of a sentence structure [30]. Moreover, it combines the simplicity of the bag-of-word models with the order-sensitivity of the sequential models. Furthermore, similar to [32], we employ a matching scheme that parallelizes interaction between each child-node in the premise text to the nodes in the hypothesis text. Where, by nodes, we mean a dependency parsed representation of the texts under consideration. Our goal is to present a simple approach with fewer parameters and that does not require exces-

---

[3] http://nlp.stanford.edu/projects/snli.

[4] Throughout the paper, we use the words 'Premise', 'Text' or 'First text' interchangeably to mean the same thing, except otherwise specified.

sive parameter optimization, while being able to generalize well. The remaining parts of the paper are organized as follows. In the next section, we give a succinct review of some related work. Then, we describe our proposed method as well as the evaluation and result.

2 Related Work

The task of recognizing textual entailment (RTE) aims at making machines to mimic human inference capability, i.e., given a text P and a ground truth hypothesis Q, humans can easily recognize whether the meaning of Q can be directly inferred from P [14]. The goal of the RTE task is to design algorithms that replicate this human capability.

Bowman et al. [9] introduced the SNLI corpus, which contains 570 k human annotated text pairs. They used a lexical classifier as a baseline for their LSTM encoding-based network. Now, their approach is simple, both the premise and hypothesis were embedded in the same space using *Glove* vectors and then the sum of vectors of words in each sentence was used as the input to the LSTM. The authors in [26] proposed a word-by-word neural attention mechanism which is also based on LSTM. LSTM seems to be the natural choice because of its ability to retain information over many time-steps, although, Convolutional Neural Networks (CNN) has likewise shown to be a good choice for similar task [34].

In the work of Rocktaschel et al. [26], two LSTMs was used. While the first LSTM is reasoning over the sequence of tokens in the text, the other is performing the same computation on the hypothesis sequence. The second LSTM is conditioned by the output of the first one, i.e., its memory is initialized by the output (i.e., the last cell state) of the last hidden state of the first LSTM when reading each input from the hypothesis. An extension of their basic model instead utilized a Bi-directional LSTM, which was used in a similar manner in order to create a dual-attention. Baudis et al. [3] also reproduced an attention model similar to the question answering model in [31]. They utilized a Recurrent Neural Networks (RNN) model, a CNN model as well as a hybrid RNN-CNN model. The RNN captures long-term dependencies and contextual representation of words before being fed to the CNN.

Parikh et al. [23] improved on the attention mechanism of [26] by introducing two components, *Compare* and *Aggregate*. The former compares aligned phrases in the premise with that of the hypothesis and vice versa, using a feed-forward neural networks. The resulting vectors are summed over in the latter component, i.e. the *Aggregate* part. Cheng et al. [12] proposed a type of LSTM with enhanced memory, called the LSTMN which is similar to the memory networks of Wetson et al. [33]. They used two attention schemes, which they called the *Shallow fusion* and the *Deep fusion*. The *Shallow fusion* considers only the attention between the words in a text, which is usually called the intra-attention. The *Deep fusion* likewise performs the intra attention but much more, it tries to identify the importance of the words in the first text, in relation to the words in the second text. This is sometimes referred to as an inter-attention. The *Deep fusion* architecture utilizes both the inter-attention and intra-attention between the text and its corresponding hypothesis. Consequently, the *Deep fusion* architecture achieved a superior performance. Overall, both models achieved near state of the art results on the tasks of textual entailment, sentiment analysis.

Wang et al. [32] introduced a model which they called the Bilateral Multi-Perspective Matching model (BiMPM) for NLI. Their model obtained the state-of-the-art result when evaluated on the SNLI corpus. BiMPM share many commonalities with a few of the previous work which are already cited above. Specifically, Wang et al. [32] also utilized a Bi-directional LSTM. Now, a Bi-directional LSTM makes use of two LSTMs that are run in parallel. The first LSTM operates on the input sequence from the first time-step to the last time-step (forward) while the second LSTM operates on the same input sequence by performing computation from the last time-step to the first time-step (backward). Essentially, the hidden state of any time-step is the concatenation of both its forward and backward hidden states. Bi-directional LSTM is thus able to capture information in both context, i.e., the past and future information.

The innovative part of the work of Wang et al. [32] is how they utilized Bi-directional LSTM (BiLSTM) for the matching scheme that they proposed. First, a BiLSTM was used to separately encode the premise and the hypothesis. Then, using any of the four matching functions that they proposed, namely, the Full matching, Maxpooling matching, Attentive matching and the Max-attentive matching, each time-step of the premise is matched against every time-steps of the hypothesis and vice-versa. Another BiLSTM then combines the result before passing through a fully connected (FC) layer for classification. Apart from the choice of matching, the architecture remains the same. They obtained the best performance when the Full-Matching function was used.

Our proposed approach also utilizes a type of LSTM, i.e., the child-sum Tree LSTM for sentence encoding. The motivation for using this is that in a dependency parsing-based tree, because a headword incorporates information from each child, a natural intra-attention is created within each sentence, since the hidden vector of the headword is composed from those of its children. Furthermore, we also incorporate a high-level interaction scheme for the two texts, using a similar matching function to the one proposed in [32]. However, our approach is more simple and has fewer parameters. The Tree-structure LSTM network builds sentence representation from headword-child subphrases of a text, but takes into account more compositional features for better generalization.

3 Methods

We describe the general LSTM architecture. Specifically, this work employs the *Child-Sum Tree*-LSTMs proposed by [30]. We describe our inter attention scheme using a form of perspective matching [32].

Long Short-Term Memory Networks

Recurrent Neural Networks (RNNs) have connections that have loops, adding feedback and memory to the networks over time. This memory allows this type of network to learn and generalize across sequences of inputs rather than individual patterns. LSTM Networks [16] are a special type of RNNs and are trained using backpropagation through time, thus overcoming the vanishing gradient problem. LSTM networks have memory blocks that are connected into layers, the block contains gates that manage the blocks state and output. These gates are the *input* gates which decides the values from the input to update the memory state, the *forget* gates which decides what information to discard from the unit and the *output* gates which decides what to output based on input and the memory of the unit. LSTMs are thus able to memorize information over a long time-steps, since this information are stored in a recurrent hidden vector which is dependent on the immediate previous hidden vector. A unit operates upon an input sequence and each gate within a unit uses the sigmoid activation function to control whether they are triggered or not, making the change of state and addition of information flowing through the unit conditional.

At each time step t, let an LSTM unit be a collection of vectors in $\mathbb{R}^d$ where d is the memory dimension: an *input gate* i_t, a *forget gate* f_t, an *output gate* o_t, a *memory cell* c_t and a *hidden state* h_t. The state of any gate can either be open or closed, represented as $[0,1]$. The LSTM transition can be represented with the following equations (x_t is the an input vector at time step t, σ represents sigmoid activation function and $\odot$ the elementwise multiplication. The u_t is a tanh layer which creates a vector of new candidate values that could be added to the state):

$$i_t = \sigma\left(W^{(i)}x_t + U^{(i)}h_{t-1} + b^{(i)}\right),$$

$$f_t = \sigma\left(W^{(f)}x_t + U^{(f)}h_{t-1} + b^{(f)}\right),$$

$$o_t = \sigma\left(W^{(o)}x_t + U^{(o)}h_{t-1} + b^{(o)}\right),$$

$$u_t = \tanh\left(W^{(u)}x_t + U^{(u)}h_{t-1} + b^{(u)}\right),$$

$$c_t = i_t \odot u_t + f_t \odot c_{t-1},$$

$$h_t = o_t \odot \tanh c_t \tag{1}$$

Tree-Structured LSTM

Tree-LSTM is a specialized type of LSTM that adopt the tree-structure topology, i.e., at any given time step t, the LSTM is able to compose its states from an input vector and hidden states of its child-nodes simultaneously. This is unlike the standard LSTM that assumes a single child per unit, since the gating vectors and memory cell updates are dependent on the states of all child-nodes. Moreover, it maintains a forget gate separately for each child node. This characteristic enables the Tree-LSTM to be able to aggregate information from each child node. A good variant of the Tree-LSTM is the *Child-Sum Tree-LSTM*, which was proposed by Tai et al. [30]. The *Child-Sum Tree-LSTM* state transition is represented by the following equations, where $C(j)$ is the set of the children of a node j, and k $\in$ $C(j)$.

$$\widehat{h}_j = \sum_{k \in C(j)} h_k,$$

$$i_j = \sigma\left(W^{(i)}x_j + U^{(i)}\widehat{h}_j + b^{(i)}\right),$$

$$f_{jk} = \sigma\left(W^{(f)}x_j + U^{(f)}\widehat{h_k} + b^{(f)}\right),$$

$$o_j = \sigma\left(W^{(o)}x_j + U^{(o)}\widehat{h}_j + b^{(o)}\right),$$

$$u_j = \tanh\left(W^{(u)}x_j + U^{(u)}\widehat{h}_j + b^{(u)}\right),$$

$$c_j = i_j \odot u_j + \sum_{k \in C(j)} f_{jk} \odot c_k,$$

$$h_j = o_j \odot \tanh(c_j) \tag{2}$$

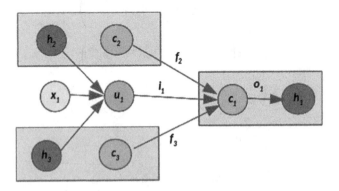

Fig. 2. Composing the memory cell c_1 and hidden state h_1 of a 2-children (subscripts 2 and 3) Tree-LSTM [30]

Tree-LSTM for Textual Entailment

We use the *Child-Sum Tree-LSTM* to generate sentence representation for both the text and hypothesis, taking as input the dependency-parse tree representation of the Premise and the Hypothesis texts. Giving a sentence, the structural connection between its constituent words form a deep branching graph, with elements and their dependencies where each connection in principle unites a head term and its dependent term(s). The dependent term maintains a one-to-one correspondence with its head, thus distinguishing between semantically useful words like nouns and verbs to say, a determinative word. With constituency parsing, a phrase-like one-to-one correspondence between the words is observed. The *Child-Sum Tree*-LSTM works better for dependency parse tree representation of a sentence, where each child is a node in the representation. For each node, the LSTM unit takes as input the vectors of its head word to which it is dependent. In the case of constituency parsing, an LSTM unit takes as input the exact vector of the node. Figure 2 shows how the hidden state and the memory cell for a Tree-LSTM unit with 2-children is composed. In the figure, labeled edges correspond to gating by the indicated gating vector, with dependencies omitted for compactness. The most important benefit of the Child-Sum Tree-LSTM is its discriminative capability, i.e., the model can learn parameters such that important words in the sentence are distinguished from unimportant words. This provides a natural and simple solution to what Attention mechanism [2] is used for.

Our approach can be divided into three parts, i.e., word encoding, sentence encoding and feature generation. Furthermore, we used two approaches for feature generation and classification, i.e., the distance based approximation and the perspective matching based approximation. The high-level representation of the two approaches is given in Figs. 3 and 4.

We assume two texts $P = (p_1,....,p_i,....,p_M)$ and $Q = (q_1,....,q_j,....,p_N)$ both of length M and N respectively. Also, a label $y \in Y$ is given, which shows the relationship, or put differently, the label for classification, e.g., entailment, neutral etc. For most of the datasets used in this work, y is either a binary output or a ternary output. P is typically the Premise or text and Q is the Hypothesis.

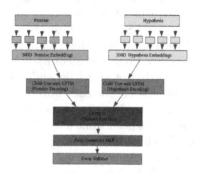

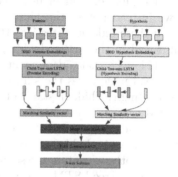

Fig. 3. A high-level illustration of the Distance-based model

Fig. 4. A high-level view of the Matching-based model

In anyway, the data representation follows the format: (P,Q,y) triples. The goal then, is to estimate the conditional probability Pr(y | P, Q) based on the training set, and predicting the relationship for testing samples by $y^* = \text{argmax}_{y \in Y}$ Pr(y | P, Q).

Irrespective of the approach, we first encode each word with a BiLSTM and then obtain a sentential representation for both the Premise and the Hypothesis using a Child-Sum Tree-LSTM, which operates on a dependency parse tree representation of the texts.

Word Encoding

Here, we represent each word in the sentences P and Q with a d-dimensional vector, where the vectors are obtained from a word embedding matrix. Generally, we make use of the 300-dimensional *GLOVE* vectors, obtained from 840 billion words [24]. A Bi-directional LSTM is then used in order to obtain contextual information between the words. A Bi-directional LSTM is essentially composed of two LSTMs, one capturing information in one direction from the first time step to the last time-step while the other captures information from the last time-step to the first. The outputs of the two LSTMs are then combined to obtain a final representation which summarizes the information of the whole sentence. Equations (3) and (4) describes this computation.

$$\overrightarrow{h_i^p} = \overrightarrow{LSTM}(\overrightarrow{h_{i-1}^p}, P_i), \quad i \in [1, ..., M]$$
$$\overleftarrow{h_i^p} = \overleftarrow{LSTM}(\overleftarrow{h_{i-1}^p}, P_i), \quad i \in [M, ..., 1] \tag{3}$$

$$h_i^f = [\overrightarrow{h_i^p}; \overleftarrow{h_i^p}] \tag{4}$$

Typically, when using an ordinary LSTM or BiLSTM to encode the words in a sentence, the whole sentence representation can be obtained as the final hidden state of the last word or time-step. The significant difference in our approach is that instead of using such final hidden state from the last time-step, we instead obtain the hidden states for each time-steps separately and then feed these hidden states into the Child-Sum Tree-LSTM. Analogously, we can also feed in the raw embedding vectors of each word into the Child-Sum Tree-LSTM, such that a dependent node takes the fixed raw vectors of its head while a head node takes the sum of its vectors and that of all its dependents. However, using the hidden states obtained when we first encode each word with a BiLSTM ensures that we have a context-aware high-level representation such that a dependent node takes the hidden state value of its head while a head node takes the sum of its hidden state along with that of all its dependents.

Consequently, we obtain the sentence representation for both P and Q, using the Child-Sum Tree-LSTM. An interesting thing is that with this setup, we do not require any form of Attention. Moreover, Attention is a way of focusing specially on some important parts of an input, and has been used extensively in some language modeling tasks [2, 23]. Essentially, it is able to identify the parts of a text that are most important to the overall meaning of the text.

Distance Based Approximation

The idea here follows the work of [30]. A high level representation of this approach is shown in Fig. 3. First, we use the child-sum Tree-LSTM to encode sentences P and Q as explained in Sect. 3, to obtain the representations h_P and h_Q.

The representations h_P and h_Q can be regarded as a high level representation of both texts P and Q. Given h_P and h_Q, we predict the label $\hat{y}_j$ by using a fully connected (FC) perceptron neural network which encodes the entailment relationship $\{r\}_j = (h_P, h_Q)$ as the distance and angle between the element-wise summed vectors of the pair (h_P, h_Q). We describe this process using Eqs. (5) and (9).

$$h_\times = h_P \odot h_Q \tag{5}$$

$$h_+ = h_P - h_Q \tag{6}$$

$$h_s = \sigma\left(W^{(\times)} h_\times + W^{(+)} h_+ + b^{(h)} \right) \tag{7}$$

$$\widehat{p}\theta(y|\{r\}_j) = softmax\left(W^{(p)} h_s + b^{(p)} \right) \tag{8}$$

$$\widehat{y}_j = \arg\max_y \widehat{p}\theta(y|\{r\}_j) \tag{9}$$

We trained our model with the negative log-likelihood of the true class labels $y^{(k)}$ according to Eq. (10), where m is the size of the training sample and is an L2 norm regularization hyperparameter.

$$J(\theta) = -\frac{1}{m} \sum_{k=1}^{m} log\widehat{p}\theta(y^{(k)}|x^{(k)}) + \frac{\lambda}{2}||\theta||_2^2 \tag{10}$$

Matching-Based Approximation

The distance based approach assumes some form of intra-attention with the child-sum Tree LSTM representation. Inter-sentence attention schemes have proven very effective at various semantic inference tasks e.g. machine translation [2] and even NLI [21]. Similar to the Distance-based approximation, we use the Child-sum Tree-LSTM to obtain the sentence representation of the premise (P) and hypothesis (Q), after separately obtaining an annotation for each word by encoding with a BiLSTM. We then use a matching-function which is similar to the one proposed by Wang et al. [32]. The matching function creates a similarity interaction between two texts, i.e., from one text to another text, this eventually creates a kind of inter-sentence attention. Note that unlike in [32], we did not include Bi-LSTM to explicitly model the contextual relationship between words of each sentence, this is already amply captured by the Child-sum Tree-LSTM which we used to encode each sentence. Also, instead of using the multi-perspective cosine function, we utilized the conventional cosine similarity without an additional trainable parameter. The matching function work as explained below.

$$\overrightarrow{match_i}^{forward} = sim(\overrightarrow{h_i}^P, \overrightarrow{h_i}^Q) \tag{11}$$

$$\overleftarrow{match_i}^{backward} = sim(\overrightarrow{h_i}^Q, \overrightarrow{h_i}^P) \tag{12}$$

$$sim = cos(V1, V2) \tag{13}$$

Given two inputs P and Q, we represent an interaction (P→Q) by a forward pass and interaction (Q→P) by the backward pass. In the forward pass (*see* Eq. 11), we compare the time-step from the last hidden state of P to every time-steps of Q. Similarly, in the backward pass (*see* Eq. 12), the computation is done in a similar way. We compare the time-step of the hypothesis from the last hidden state of Q to each of the time-steps in P. For both forward and backward passes, the comparison is done by obtaining how similar the two vectors are, using the cosine similarity formula in Eq. (13). This matching function creates a form of interconnection from one-time-step to every other time-steps, thus yielding two vectors of similarity scores.

In the original full-matching method of [32], they compared each time-step from one text to every time-step in the other text. Furthermore, the comparison is done with a Bi-LSTM which makes the approach further computationally expensive. Here, we only compare the sentence representation of one sentence with each word in the other sentence and vice-versa. Also, for simplicity, we use the hidden state from the last time-step of a text as its encoding representation.

Once we obtain the similarity score vectors, i.e., S_p and S_q for P and Q respectively, we introduce a merge layer in order to concatenate the two vectors. The resulting vector is then passed to a fully connected Multilayer Perceptron (MLP) network to learn the entailment relationship. The predicted class is obtained from the probability distribution given in Eq. (14). In order to train our neural network, we use Multi-Class Cross-Entropy loss function, with 20% dropout regularization [29].

$$\hat{y} = H([s_p; s_q]) \tag{14}$$

$$\hat{y} = \arg\max_y y_{(i)} | x_{(i)} \tag{15}$$

We trained our model with the cross-entropy loss given in Eq. 16 where θ_F, θ_G, θ_H are parameters to be learned.

$$L(\theta_F, \theta_G, \theta_H) = \frac{1}{J} \sum_{j=1}^{J} \sum_{c=1}^{C} y_c^{(j)} \log \frac{\exp(\hat{y}_c)}{\sum_{c=1}^{C} \exp(\hat{y}_c)} \tag{16}$$

4   Evaluation

The RTE PASCAL challenge [14] is an important avenue for researchers to submit TE systems for public evaluation. We evaluated our system on the PASCAL RTE3 dataset which consists of 800 sentence pairs both for development and test set. The RTE3 dataset has only two classes, i.e., the entailment relation can either be true or false. The SEMEVAL track offering similarity and entailment tasks also make use of the SICK dataset[5]. SICK consists of 10000 sentence pairs annotated for use in both sentence similarity and 3-way entailment task. Finally we evaluated our system on the SNLI corpus [9] which is a big entailment dataset that is publicly available.

In the context of our ongoing work in the legal domain[6] [5, 7, 8], we evaluated our models on a legal dataset of textual entailment. The three datasets cited above contain sentences that are domain independent and thus have no technical jargons. Our goal is to see how our model would perform within the complex legal domain. Legal texts seem intuitive, because they have some peculiarities which set them apart from day-to-day texts, since they employ legislative terms. For instance, a sentence can have a reference to another sentence (e.g., an article) without any explicit link to its text from within the quoting text. Also, sentences are usually long with several clausal dependencies, that is notwithstanding of its inter and intra-sentential anaphora resolution complexity. We opined that a system that is able to achieve good result in this scenario would generalize well given other domain dependent texts.

We used the COLIEE dataset[7] which is a Question-Answering legal corpus made available in the context of COLIEE Legal Information Extraction/Retrieval challenge. Task 2 of the challenge addresses NLI task, such that, given a sentence and a Query, a system identifies if there is an entailment or not. We provide our evaluation result on the 2015 training and test sets.

Experiment

In this work, we used the Child-sum Tree-LSTM similar to the one proposed in [30] to encode the texts. We obtained dependency tree of both the text and the hypothesis using Stanford dependency parser [10]. To embed the training data, We used 300-dimensional Glove vectors [24]. Also, we keep the weights of the embeddings fixed and thus, not trainable.

Our model was implemented based on Keras[8]. For the COLIEE, SICK and RTE task, we used sigmoid for distributing output probability while we used softmax for SNLI with three classes. We apply a random *Dropout* of 0.2 throughout the models.

---

[5] http://clic.cimec.unitn.it/composes/sick.html.
[6] Specifically, the MIREL project: http://www.mirelproject.eu, which is drawn from our past project EUCases [6].
[7] http://webdocs.cs.ualberta.ca/~miyoung2/COLIEE2016.
[8] https://github.com/fchollet/keras.

We used a uniform batch-size of 25 in all the experiments excluding that of SNLI dataset, with the batch-size = 256. We used ADAM, a stochastic optimizer with learning rate set at 0.01 and a decay value of 1e-4. Moreover, we used early-stopping, in order to keep track of the point where the loss ceases to decrease after 4 epochs. This also helps to reduce over-fitting on the training set.

Tables 1, 2, 3 and 4 shows the evaluation result on the datasets that we used in our experiment. For Table 2, we used the results from Rocktaschel et al., [26], Baudis et al. [3] and Bowman et al. [9] as the baseline systems on SNLI and SICK respectively. For PASCAL-RTE3, we compare our models to some ML classifier baselines since there is no recent work which use similar deep learning approach on that dataset. For the result on COLIEE dataset in Table 4, we include the result reported by [18] on the same dataset. We can see that the performance of our model is near state-of-the-art, even though we slightly have fewer trainable parameters when compared with some of the baseline systems. Our model that is based on the matching-interaction approach seems to generally outperform our second model. The performance on SNLI corpus is very close to that of Wang et al. [32], even though the architecture of our model is simpler in theory.

Table 1. Evaluation on SNLI dataset

Model	k	$-\theta-_m$	Train	Test
LSTM [9]	100	220 k	84.8	77.6
Classifier [9]	-	-	**99.7**	78.2
Neural Attention [26]	100	250 k	85.3	83.5
NTI-SLSTM-LSTM encoders [22]	300	400 k	82.5	83.4
BiLSTM encoders with intra-attention [21]	600	2.8 m	85.9	85.0
LSTMN with deep attention fusion [12]	450	3.4 m	88.5	86.3
ESIM + 300D Syntactic TreeLSTM [11]	600	7.7 m	93.5	88.6
BiMPM Ensemble [32]	300	6.4 m	93.2	**88.8**
Tree-LSTM-Distance-Angle (**This Paper**)	300	560 k	87.3	84.1
Tree-LSTM-Matching (**This Paper**)	300	2.2 m	90.6	86.4

Table 2. Evaluation on SICK dataset

Model	SICK Train	SICK Test
attn1511 [3]	85.80	76.70
LSTM-RNN [9]	**99.90**	80.80
Tree-LSTM-Distance-Angle (**This Paper**)	85.10	76.00
Tree-LSTM-Full-Matching (**This Paper**)	95.60	**81.80**

Table 3. Evaluation on PASCAL-RTE3 dataset

Model	Train	Test
Wikipedia co-training [35]	-	57.25
Wiki + RTE-3 [35]	-	59.00
SVM [25]	-	66.37
Naive Bayes [25]	-	65.87
Sha et al. [27]	-	85.16
Tree-LSTM-Distance-Angle	**95.76**	86.90
Tree-LSTM-Full-Matching	93.80	**89.20**

Table 4. Evaluation on COLIEE 2015 Dataset

Model	Accuracy (%)
Convolutional Neural Network [19]	51.5
Convolutional Neural Network with TF-IDF [19]	53.0
Convolutional Neural Network with LSA [19]	54.5
Tree-LSTM-Distance-Angle (**This Paper**)	53.84
Tree-LSTM-Full-Matching (**This Paper**)	**57.40**

Discussion and Conclusions

In this paper, we described a *Child-Sum Tree*-LSTM model which obtained a good result on 3 well known textual entailment datasets. The results of our evaluation are given in Tables 1, 2, 3, and 4.

We noticed that the worst result was obtained on the COLIEE corpus. However, the COLIEE corpus, with less than 500 training samples may not be a good corpus for neural networks. Also, the uniqueness of the corpus, being legislative texts, may also be a factor. Furthermore, compared to SNLI and SICK, both the text and hypothesis in COLIEE are unusually long. Nevertheless, we report improved result to the baselines from [18].

In Table 1, we can see that our result is very close to the current state-of-the-art system [32], which has more than double trainable parameters, compared to ours. In Table 2, we see that we obtained the best result on the SICK dataset. Even though the RTE-3 dataset contains small training samples, still our models significantly outperformed the ML classifiers, i.e., SVM and Naive Bayes. In all the experiments, we obtained better accuracy with our interaction or matching-based approximation model than the one with the distance-based approximation model. This is because of the fact that entailment is not just a function of similarity between the texts being compared as in the case of the distance-based approximation model. Compared to all the benchmarked baselines, our approach is simple, requires no attention mechanism and has fewer trainable parameters. In the future, we would like to do a comparative qualitative analysis of our two models. Also, we would like to see if incorporating attention will impact the performance significantly.

Acknowledgments. Kolawole J. Adebayo has received funding from the Erasmus Mundus Joint International Doctoral (Ph.D.) programme in Law, Science and Technology. Luigi Di Caro and Guido Boella have received funding from the European Union's H2020 research and innovation programme under the grant agreement No 690974 for the project "MIREL: MIning and REasoning with Legal texts". Livio Robaldo has received funding from the European Union's H2020 research and innovation programme under the grant agreement No 661007 for the project "ProLeMAS: PROcessing LEgal language in normative Multi-Agent Systems".

References

1. Androutsopoulos, I., Malakasiotis, P.: A survey of paraphrasing and textual entailment methods. J. Artif. Intell. Res. **38**, 135–187 (2010)
2. Bahdanau, D., Cho, K., Bengio, Y.: Neural machine translation by jointly learning to align and translate. arXiv preprint arXiv:1409.0473 (2014)
3. Baudiš, P., Šedivỳ, J.: Sentence pair scoring: towards unified framework for text comprehension. arXiv preprint arXiv:1603.06127 (2016)
4. Bentivogli, L., Clark, P., Dagan, I., Giampiccolo, D.: The seventh pascal recognizing textual entailment challenge. In: Proceedings of TAC 2011 (2011)
5. Boella, G., Di Caro, L., Humphreys, L., Robaldo, L., Rossi, R., van der Torre, L.: Eunomos, a legal document and knowledge management system for the web to provide relevant, reliable and up-to-date information on the law. In: Artificial Intelligence and Law (2016, to appear)
6. Boella, G., Di Caro, L., Graziadei, M., Cupi, L., Salaroglio, C.E., Humphreys, L., Konstantinov, H., Marko, K., Robaldo, L., Ruffini, C. and Simov, K., Violato, A., Stroetmann, V.: Linking legal open data: Breaking the accessibility and language barrier in european legislation and case law. In: Proceedings of the 15th International Conference on Artificial Intelligence and Law, ICAIL 2015. ACM, New York (2015)
7. Boella, G., Di Caro, L., Rispoli, D., Robaldo, L.: A system for classifying multi-label text into EuroVoc. In: Proceedings of the Fourteenth International Conference on Artificial Intelligence and Law, ICAIL 2013, pp. 239–240. ACM, New York (2013)
8. Boella, G., Di Caro, L., Robaldo, L.: Semantic relation extraction from legislative text using generalized syntactic dependencies and support vector machines. In: Morgenstern, L., Stefaneas, P., Lévy, F., Wyner, A., Paschke, A. (eds.) RuleML 2013. LNCS, vol. 8035, pp. 218–225. Springer, Heidelberg (2013). doi:10.1007/978-3-642-39617-5_20
9. Bowman, S. R., Angeli, G., Potts, C., Manning, C.D.: A large annotated corpus for learning natural language inference. arXiv preprint arXiv:1508.05326 (2015)
10. Chen, D., Manning, C.D.: A fast and accurate dependency parser using neural networks. In: EMNLP, pp. 740–750 (2014)
11. Chen, Q., Zhu, X., Ling, Z., Wei, S., Jiang, H.: Enhancing and combining sequential and tree lstm for natural language inference. arXiv preprint arXiv:1609.06038 (2016)
12. Cheng, J., Dong, L., Lapata, M.: Long short-term memory-networks for machine reading. arXiv preprint arXiv:1601.06733 (2016)
13. Dagan, I., Dolan, B., Magnini, B., Roth, D.: Recognizing textual entailment: rational, evaluation and approaches-erratum. Nat. Lang. Eng. **16**(01), 105–105 (2010)
14. Dagan, I., Glickman, O., Magnini, B.: The PASCAL recognising textual entailment challenge. In: Quiñonero-Candela, J., Dagan, I., Magnini, B., d'Alché-Buc, F. (eds.) MLCW 2005. LNCS, vol. 3944, pp. 177–190. Springer, Heidelberg (2006). doi:10.1007/11736790_9
15. Feng, M., Xiang, B., Glass, M.R., Wang, L., Zhou, B.: Applying deep learning to answer selection: a study and an open task. In: 2015 IEEE Workshop on Automatic Speech Recognition and Understanding (ASRU), pp. 813–820. IEEE (2015)
16. Hochreiter, S., Schmidhuber, J.: Long short-term memory. Neural Comput. **9**(8), 1735–1780 (1997)
17. John, A.K., Di Caro, L., Boella, G.: Normas at semeval-2016 task 1: Semsim: a multi-feature approach to semantic text similarity. In: Proceedings of SemEval (2016)

18. Kim, M.Y., Xu, Y., Goebel, R.: A convolutional neural network in legal question answering (2015)
19. Kim, M.-Y., Xu, Y., Goebel, R.: Legal question answering using ranking SVM and syntactic/semantic similarity. In: Murata, T., Mineshima, K., Bekki, D. (eds.) JSAI-isAI 2014. LNCS, vol. 9067, pp. 244–258. Springer, Heidelberg (2015). doi:10.1007/978-3-662-48119-6_18
20. Liu, P., Qiu, X., Huang, X.: Modelling interaction of sentence pair with coupled-LSTMs. arXiv preprint arXiv:1605.05573 (2016)
21. Liu, Y., Sun, C., Lin, L., Wang, X.: Learning natural language inference using bidirectional LSTM model and inner-attention. arXiv preprint arXiv:1605.09090 (2016)
22. Munkhdalai, T., Yu, H.: Neural semantic encoders. arXiv preprint arXiv:1607.04315 (2016)
23. Parikh, A.P., Täckström, O., Das, D., Uszkoreit, J.: A decomposable attention model for natural language inference. arXiv preprint arXiv:1606.01933 (2016)
24. Pennington, J., Socher, R., Manning, C.D.: Glove: global vectors for word representation. In: EMNLP, vol. 14, pp. 1532–1543 (2014)
25. Gaona, M.A.R., Gelbukh, A., Bandyopadhyay, S.: Recognizing textual entailment using a machine learning approach. In: Advances in Soft Computing, pp. 177–185 (2010)
26. Rocktäschel, T., Grefenstette, E., Hermann, K.M., Kočiský, T., Blunsom, P.: Reasoning about entailment with neural attention. arXiv preprint arXiv:1509.06664 (2015)
27. Sha, L., Li, S., Chang, B., Sui, Z., Jiang, T.: Recognizing textual entailment using probabilistic inference. In: EMNLP, pp. 1620–1625 (2015)
28. Socher, R., Huang, E., Pennin, J., Manning, C., Ng, A.: Dynamic pooling and unfolding recursive autoencoders for paraphrase detection. In: Advances in Neural Information Processing Systems, pp. 801–809 (2011)
29. Srivastava, N., Hinton, G.E., Krizhevsky, A., Sutskever, I., Salakhutdinov, R.: Dropout: a simple way to prevent neural networks from overfitting. J. Mach. Learn. Res. 15(1), 1929–1958 (2014)
30. Tai, K.S., Socher, R., Manning, C.D.: Improved semantic representations from tree-structured long short-term memory networks. arXiv preprint arXiv:1503.00075 (2015)
31. Tan, M., Xiang, B., Zhou, B.: LSTM-based deep learning models for non-factoid answer selection. arXiv preprint arXiv:1511.04108 (2015)
32. Wang, Z., Hamza, W., Florian, R.: Bilateral multi-perspective matching for natural language sentences. arXiv preprint arXiv:1702.03814 (2017)
33. Weston, J., Chopra, S., Bordes, A.: Memory networks. arXiv preprint arXiv:1410.3916 (2014)
34. Yin, W., Schütze, H., Xiang, B., Zhou, B.: Abcnn: attention-based convolutional neural network for modeling sentence pairs. arXiv preprint arXiv:1512.05193 (2015)
35. Zanzotto, F.M., Pennacchiotti, M.: Expanding textual entailment corpora from wikipedia using co-training. In: Proceedings of the COLING-Workshop on The Peoples Web Meets NLP: Collaboratively Constructed Semantic Resources, vol. 128 (2010)

Extracting Core Claims from Scientific Articles

Tom Jansen[✉] and Tobias Kuhn

Faculty of Science, Vrije Universiteit Amsterdam,
De Boelelaan 1105, 1081 HV Amsterdam, The Netherlands
t.d.jansen@student.vu.nl, t.kuhn@vu.nl

Abstract. The number of scientific articles has grown rapidly over the years and there are no signs that this growth will slow down in the near future. Because of this, it becomes increasingly difficult to keep up with the latest developments in a scientific field. To address this problem, we present here an approach to help researchers learn about the latest developments and findings by extracting in a normalized form core claims from scientific articles. This normalized representation is a controlled natural language of English sentences called AIDA, which has been proposed in previous work as a method to formally structure and organize scientific findings and discourse. We show how such AIDA sentences can be automatically extracted by detecting the core claim of an article, checking for AIDA compliance, and – if necessary – transforming it into a compliant sentence. While our algorithm is still far from perfect, our results indicate that the different steps are feasible and they support the claim that AIDA sentences might be a promising approach to improve scientific communication in the future.

Keywords: Core claims · Core sentences · AIDA · Text mining · Information extraction · Scientific findings

1 Introduction

The number of scientific articles is rapidly growing [8]. With this overwhelming amount of information, the need for proper information extraction tools that only extract the relevant information is increasing too. There is so much textual information out there that scientists can easily get lost looking for the right information. This overload makes it hard and time-consuming for researchers to keep up with the latest developments in their scientific field. When finding the latest developments proves too difficult, duplicate research may be conducted leading to unnecessary work and research that is already conducted elsewhere. The field of text mining [1, 17] has played a crucial role in extracting relevant information from (scientific) literature. Text mining is primarily used for the decomposition and analysis of texts or textual information. Generally, it refers to the process of extracting interesting – or relevant – information or patterns from unstructured documents [17]. Often used techniques include part-of-speech tagging, named entity recognition and sentiment analysis. Techniques like these pave the way for more complex and interesting text analyses. The extraction of information and knowledge from texts is such a complex step. Over the years, enormous progress has been made with respect to text mining in the areas of information retrieval, evaluation methodologies and resource construction especially in the biological domain [21].

© Springer International Publishing AG 2017
T. Bosse and B. Bredeweg (Eds.): BNAIC 2016, CCIS 765, pp. 32–46, 2017.
DOI: 10.1007/978-3-319-67468-1_3

Despite this progress, the problem of information overload has not been solved yet, and it seems that we need more than just text mining. To overcome the issue at hand and improve the way we share scientific findings, we might have to change scientific communication altogether. A possible solution would be that text mining and other machine learning approaches are supported by the way we publish scientific findings in the future. The concept of AIDA sentences is such an approach and offers a way to structure scientific findings and discourse [7]. AIDA sentences are a controlled natural language [6] of single independent English sentences that can represent scientific claims in a normalized and re-usable fashion. AIDA sentences are proposed as a tool for researchers to easily access and communicate research hypotheses, claims and opinions. The vision is that scientists will summarize their findings in AIDA sentences and interlink them with other claims, but to take into account the large body of existing literature, we need to apply text mining methods to integrate past and future research in a symbiosis of manual and automated work. In this paper, an attempt is made to start this integration of past and future research. To do this, we present a 3-fold approach to extract core claims from scientific articles to bootstrap the AIDA approach. Step one is to find the core sentence of a scientific article. Second, a check is performed to see whether the sentence complies with the AIDA rules. If it does not, a third and final step is undertaken that attempts to rewrite the core sentence into an AIDA sentence.

The rest of this paper is organized as follows. Section 2 presents background information and related work that may be helpful in achieving the goal of this paper. Here, the concept of AIDA sentences is explained in detail as well. Section 3 provides a more elaborate explanation of the approach. In Sect. 4 the experimental framework that is used to train and test the developed algorithm is discussed. Results of these tests are provided in Sect. 5, after which they are discussed in Sect. 6. Finally, the document is concluded in Sect. 7.

2   Background

Below, we introduce the concept of AIDA sentences, which plays a key role in our approach. Therefore, this will be addressed first. Closely related to the goal of this paper, and widely researched, is the topic of document summarization. This will be addressed here as well. Before we delve into document summarization, however, we take a look at related work with respect to keyword and keyphrase extraction – which is often an important step in document summarization.

2.1   AIDA Concept

AIDA sentences have been proposed as a tool for researchers to access and communicate scientific claims [7]. An AIDA sentence represents a single scientific claim that provides the core finding of an article. When such claims are extracted from articles and easily accessible, researchers can easily keep up with the latest developments. There are four requirements that need to be fulfilled by a sentence to be AIDA – each denoted by a single letter in AIDA. This structure will play a key role in this paper and its approach to extract scientific claims from articles. Here, we show two example sentences that comply with all the requirements to be AIDA sentences:

- Malaria is transmitted by mosquitos.
- The degree of hepatic reticuloendothelial function impairment does not differ between cirrhotic patients with and without previous history of SBP.

These sentences not only comply with all the requirements, they also display a clear statement of the scientific finding of a given article – embracing the bigger picture of AIDA sentences. In other words; AIDA sentences comply with a set of requirements, which make sure the sentence provides a single scientific claim in a normalized fashion. Ideally, such a sentence is written by the authors themselves. However, since this is often not enforceable (in particular for already written articles), they can be automatically extracted to get it started. The AIDA requirements, quoted from [7], are:

- **Atomic:** a sentence describing one thought that cannot be further broken down in a practical way
- **Independent:** a sentence that can stand on its own, without external references like "this effect" or "we"
- **Declarative:** a complete sentence ending with a full stop that could in theory be either true or false
- **Absolute:** a sentence describing the core of a claim ignoring the (un)certainty about its truth and ignoring how it was discovered (no "probably" or "evaluation showed that"); typically in present tense

The absoluteness criterion might sound suspicious at first, as uncertainty and the context of discoveries are crucial aspects of scientific findings. In fact, the AIDA approach does not deny the importance of these aspects, but assumes that they can be formally linked to an AIDA sentences, using models such as the ORCA model [19]. Therefore, it is argued that these aspects do not need to be present in the sentences. Together, these requirements should allow AIDA sentences to be re-used (e.g. another paper reproducing a claim) and interlinked (subsumption hierarchies, equivalence/relatedness, etc.), and thereby enable the efficient organization of scientific discourse. Furthermore, formal provenance and metadata (including scientific method, degree of confidence, source article, authors, timestamp, used data, etc.) can be attached to AIDA sentences with the nanopublication technique [11].

2.2   Keyword and Keyphrase Extraction

Keywords are often defined as the most important words of a scientific article. They comprise the subject(s) of an article and provide a concise summarization of a document. Because of this, they are an important feature for techniques such as document retrieval and topic search [20]. Whereas keywords are single word terms, keyphrases consist of multiple words (e.g. *keyword extraction*). Authors often include manually assigned keywords/keyphrases in their articles, but there are still many articles that lack any of the two. There are a number of approaches by which the extraction of keywords and keyphrases can be carried out. Broadly speaking, there are four methods: rule-based linguistic approaches, statistical approaches, machine learning approaches (supervised, unsupervised and semi-supervised) and domain specific approaches [16]. An – old-fashioned but still effective – statistical approach that is often used is TF-IDF. This feature is used to reflect the importance of a word to a document in a collection or corpus. TF-IDF stands for Term Frequency-Inverse Document Frequency and determines the importance based on the frequency of a word in a document versus the frequency of that word in the whole corpus. This simple algorithm efficiently categorizes relevant words [12]. Of the four methods, however, machine learning approaches are most prevalent in academic literature [3]. Early research into keyphrase extraction approaches this problem as a supervised learning task [18]. In that research, a genetic algorithm and a set of adjustable parameters were used for keyphrase extraction. A more recent study proposed an extended Term Frequency method to extract keywords [5]. Multiple features were used and were given weighting methods, after which an SVM model was used on the results for further optimization. In an unsupervised approach, another study extracts keyphrases based on the observation that keyphrases frequently contain multiple words but rarely standard punctuation or stop words [13]. Finally, research has shown that the highest frequency of keywords per noun in biomedical texts is found in the abstract [15]. Since the abstract is a short summary of the article, it makes sense this part contains a great deal of relevant information in a relatively small piece of text. However, it is also stated that other sections of biomedical texts are worthwhile to go through since they potentially host many keywords as well.

2.3   Document Summarization

Like keyphrase extraction, automatic document summarization has received a lot of attention over the years. On the one hand, both are similar in the fact that they aim to determine the essence of a text or document. On the other hand, document summarization is more complex because it not only deals with words or phrases but whole sentences and larger bodies of text. There are two main methods of automatic summarization: extractive and abstractive. An extractive summary contains a set of sentences from the document, whereas an abstractive summary can contain material that is not present in the document but is constructed [2, 14]. A popular approach to summarize documents nowadays is based on graph representations. TextRank is a well-known graph-based ranking model that is used for both keyword and sentence extraction [10]. The importance of words and sentences is based on the relation between them within the constructed graph. In the graph, sentences are represented as vertices. The higher the number of relations of a vertex within the graph, the higher the importance of that vertex. Another recent research provides a comparison between an extractive and abstractive approach to document summarization [9]. The extractive approach consists of five steps of which the middle step is topic identification. A summary is generated containing sentences from the document that are considered most relevant. For the abstractive approach, the extractive summary is used as a basis and a word graph is generated that is integrated with that summary to create an abstractive summary. Results show that both approaches perform similar in terms of the information the summaries contain, but that the abstractive summary is more appropriate from a human perspective. Finally, a study performed a quantitative and qualitative assessment of 15 algorithms for sentence scoring [4]. Out of the 15 assessed algorithms, five methods showed the best performance: word frequency, TF-IDF, lexical similarity, sentence length and the TextRank score.

3   Approach

To achieve the goal of extracting a single claim from a scientific article, we employ a rule-based approach. While machine learning approaches, as introduced in the Background section, have shown to perform well for a variety of problems, they can be expected to require training data of considerable size to lead to satisfactory results for our problem domain. Such datasets are not yet present with respect to scientific articles and their core sentences. A rule-based approach is therefore a natural first approach that could facilitate the creation of training data to employ machine learning methods on this problem in the future. Our rule-based approach consists of three steps: (1) the core sentence of an article is extracted; (2) that sentence is checked for AIDA compliance; (3) when the sentence does not comply with the requirements for an AIDA sentence, an attempt is made to rewrite the sentence to satisfy the requirements. At the highest level, the process of retrieving a scientific claim from an article is shown in Fig. 1.

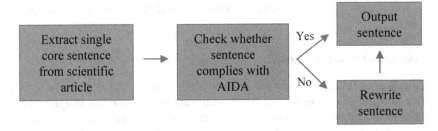

Fig. 1. Process of retrieving a scientific claim.

3.1   Extracting Core Sentences

First, the core sentence needs to be extracted from a scientific article. Core sentences are sentences that best describe the core idea or claim of an article. For the scope of this research, we have limited the search for the core sentence to the abstract of scientific articles alone. As denoted earlier, research has shown that the abstract contains the highest frequency of keywords for biomedical texts. To set up this research, we have therefore chosen to focus on the abstract. Furthermore, the assumption is made that a scientific article – or abstract in our case – contains one single sentence that best describes the core claim of that article. To find the core sentence, every sentence is rated with a score based on its content. When every sentence has received a score, the sentence with the highest score is extracted as the core sentence. The score of a sentence is based on four factors that will be briefly explained:

1. Whether the sentence matches a defined pattern
2. The number of 'core' and 'non-core words' in the sentence
3. Term frequency
4. Sentence length

Core sentences of scientific articles often start with a summarizing phrase like *"overall, this study reveals that"*, *"in conclusion, these findings confirm"*, *"the data suggest that"* etc. If an article contains a sentence with a similar structure or pattern, the chance is relatively high that this is the core sentence. To find sentences like these, we define these patterns by a regular expression. If a sentence is found that matches the regular expression, this sentence gets a relatively high score. The search continues, since several sentences can be structured like this.

Although not every sentence is structured in the (desired) way that is shown in the regular expression, they usually still contain words of the same scope. Verbs that are also used in the former regular expression like *"reveal"*, *"confirm"*, *"suggest"* potentially show that a scientific claim is coming. This also holds for nouns like *"study"*, *"experiment"* and *"research"* and adjectives like *"overall"*, *"altogether"* and *"collectively"*. We stored all these and other similar words into a list. If a word from this list is present in a sentence, the score of the sentence is increased. Another list of words is also created, containing words that should not be present in a core sentence. These are words like *"sample"*, *"survey"* and *"interview"*. When one of these words is present, the score of the sentence is decreased.

Third, our algorithm creates a list that contains the ten most frequent terms of the article. As denoted in the Background section, term frequency and TF-IDF are often used and have proven to be successful. For this study, we have chosen to use term frequency alone. A list of the ten most frequent terms is compiled – excluding English stop words. For this particular step, the algorithm does not limit itself to the abstract but looks at the entire article to search for the most frequent words. Our algorithm assumes that words from this list are most important for the current article and therefore have a high chance of forming the claim of that article. When a frequent word is found, the score of the sentence in which the word is present is increased.

Finally, the length of a sentence is used as an indicator as well. Ideally, a core claim is not too long consisting only of a simple claim – like the example sentences shown above. In abstracts and articles, however, core sentences can be relatively long. To avoid very long sentences to be extracted, for example because they do contain a lot of important words, we penalize sentences that are longer than any of the labeled core sentences plus ten percent.

3.2 Checking AIDA Rules

The second step of the algorithm is to check whether the extracted core sentence complies with the AIDA rules. For every individual rule, several checks are conducted. For three of the four rules (Atomic, Independent and Absolute) the most important check consists of a list of words that provide a strong indication for a violation of that particular rule.

An atomic sentence describes a single thought. We compiled a list of words that indicate, for example, a contradistinction or enumeration – implying more than one thought. In addition to this list, the algorithm checks whether the sentence consists of one or multiple clauses. Every sentence is parsed using a statistical parser. From the parsed tree, the algorithm derives whether the sentence consists of a single clause or several. If multiple clauses are present, the sentence is considered not atomic because it can potentially be broken down further.

A sentence is independent when there are no external references and the sentence can stand on its own. Our algorithm uses a list of phrases such as *"this study"*, *"this experiment"* and *"we"* to check for external references. When such a word is present, it means the author refers to something else than just the current sentence and therefore the sentence is not independent.

A sentence needs to be grammatically complete and correct and in theory verifiable to be declarative. For this requirement, the algorithm checks whether the sentence is grammatically well-formed. It checks whether it starts with a capital and ends with a full stop. To be grammatically correct a sentence needs to contain at least one verb and one noun. If any of these rules is violated, our algorithm flags the sentence as not declarative.

An absolute sentence describes the core of a claim without showing how that claim was discovered and whether its truth is (un)certain. The list of words for this rule contains words like *"probably"*, *"likely"*, *"suggest"* since they show (un)certainty. Furthermore, our algorithm looks for modal verbs in the sentence. Modal verbs are used to indicate modality – that is: certainty, probability, possibility, permission and so on. When a modal verb is present or any word in the compiled list, the sentence is considered not absolute. Finally, claims that are absolute are written in the present tense. The algorithm therefore looks for verbs that are written down in the past tense. When a verb in the past tense is found, the algorithm identifies the sentence as not absolute.

3.3 Rewriting Non-AIDA Sentences

When the core sentence is extracted and the outcome of the AIDA check is negative, the algorithm attempts to rewrite that sentence into one that does comply with the AIDA rules.

Our current algorithm does not cover the cases where an extracted sentence is not atomic or not independent. Rewriting these sentences are the most challenging cases, and our current study focuses on the low-hanging fruits. If a sentence is not declarative, however, the algorithm ensures that the rewritten sentence will start with a capital and ends with a full stop. When the extracted sentence is not absolute, a number of steps are taken. First of all, a regular expression is used to find and remove expressions like *"overall, the results show that"* and *"these findings indicate"*. Second, our algorithm looks for modal verbs and the verb that comes with this modal verb. These are then replaced with a single verb without modality. Verbs in past tense are dealt with in a similar fashion. In this case, the algorithm identifies whether the corresponding noun is in singular or plural form. When the form of the noun is retrieved, the verb is rewritten into the present tense. Finally, we remove words showing (un)certainty – using the same list we use to check for absoluteness.

After the requirements are processed, a final syntactic check is done to deal with double spaces and to ensure that all sentences start with a capital letter and end with a full stop.

4   Experimental Framework

Here, we provide a description of the framework that we used to train and test our algorithm. First, we will discuss the used modules and describe our data set. Second, the experiments that we conducted to train and test the algorithm are explained. The three steps of our algorithm that were specified in the previous section, are evaluated and therefore also described individually.

4.1   Modules and Data Set

Python 2.7 in combination with the NLTK library[1] is used for Natural Language Processing. NLTK modules are used to separate sentences, tokenize the sentences and to assign a Part-Of-Speech tag to every token. Furthermore, we use a statistical parser to parse sentences: the pyStatParser[2].

For our experiments, we make use of a total set of 250 articles from the PubMed Central FTP service[3]. From this FTP service, we took three archives (comm_use.0-9A-B.txt.tar, comm_use.C-H.txt.tar and comm_use.I-N.txt.tar). Out of these three archives, 250 articles were randomly picked from a wide variety of journals. These 250 articles were divided into a training set and a test set. The training set consists of 125 articles from six different journals. For the test set we ensured that there are no more than three articles of the same journal. This, to ensure that our algorithm would not be trained or tested domain specific but work equally well for all scientific domains. For every article, we handpicked the core sentence of the abstract beforehand to create our gold standard. In addition to these 250 random articles, we also used a list of 250 already checked unique sentences, taken from the result table[4] of existing work on AIDA sentences [7]. Out of the 250 checked sentences, 65 sentences were labeled as not compliant with all of the AIDA rules.

The full code and other required files, including the gold standards and comprehensive result files, can be found on GitHub[5].

4.2   Experimental Setup

Here, we describe for each individual step of our approach, how we trained and tested the algorithm.

The first step that is trained and tested is that of extracting core sentences. As mentioned earlier, for this research we focus on extracting the sentence from the abstract alone. Our algorithm uses a number of parameters to give sentences a score. These parameters were optimized using the training set of 125 random articles. All the

---

[1] http://www.nltk.org/.

[2] https://github.com/emilmont/pyStatParser - accessed on 09/03/2017.

[3] https://ftp://ftp.ncbi.nlm.nih.gov/pub/pmc/oa_bulk/ - accessed on 10/03/2017.

[4] https://github.com/tkuhn/nanopubstudies-supplementary/blob/master/botstudy/extract_evalresults. ods - accessed on 15/03/2017.

[5] https://github.com/TomJansen25/Extracting-Core-Claims/.

parameters either increase or decrease the score of a sentence with a maximum of 50 points. During optimization, many different combinations were tried before ending up with the optimal configuration. After this, testing was done with our test set of 125 different random articles. We then compared the 125 extracted sentences from the test set with the created gold standard to assess performance.

Second, we check whether sentences comply with the AIDA requirements. We trained our algorithm using the 250 sentences from the table from existing work on AIDA sentences. To measure the performance of the algorithm we used the 125 sentences that our algorithm extracted from the used articles. Before they were checked by the algorithm, we created another gold standard by checking the sentences ourselves. After we labeled all the sentences, we ran the algorithm and compared the results for evaluation.

Finally, the last step of our algorithm is to rewrite the sentences that do not comply with the AIDA requirements. For this step, we used the 250 entries from existing work on AIDA sentences for training again. During training, the sentences that were not compliant with all the requirements were rewritten and manually evaluated to improve the algorithm. After training, we tested this part using the correctly extracted sentences from the 125 abstracts that are used in the previous two steps. Given that the algorithm is trained to rewrite core sentences into single claims, we have chosen to omit the extracted sentences that did not match our gold standard and are thus not a core sentence. To assess the performance of this step, we manually evaluated the rewritten sentences.

5   Results

Below, we show the performance of the individual steps of our algorithm.

5.1   Extracting Core Sentences

Before running the algorithm on the abstracts of 125 random articles, all the core sentences were handpicked. Due to the use of plain text files from the PubMed Central archives, in some cases the algorithm extracts some formatting (mostly a header) along with the sentence. To accommodate this, we compare the extracted sentence with the labeled core sentence from our gold standard and check how similar they are. When they show a similarity of at least 85%, the extracted core sentence is considered similar enough to the labeled core sentence. This is then a correctly extracted core sentence. When extracting the core sentence from the abstract, 77 out of 125 sentences match the labeled core sentence from the gold standard. Furthermore, in the remaining 48 sentences, ten extracted sentences also showed a claim of the article but not the core claim. Overall, 87 sentences containing claims were extracted from 125 abstracts of which 77 were also labeled beforehand as the core sentence. Thus, in 69.6% of the cases, a sentence is extracted from the abstract that contains a claim and in 61.6% of the cases, the perfect core sentence is extracted from the abstract.

5.2   Checking Sentences for AIDA Compliance

The set of 125 extracted sentences from the previous step is used to test the checks for AIDA compliance. The performance of the algorithm is measured using a set of four metrics: precision, recall, F-measure and accuracy. The diagram below (Fig. 2) shows an overview of the four metrics for every individual check and the AIDA check as a whole. Salient numbers will be discussed in the next section.

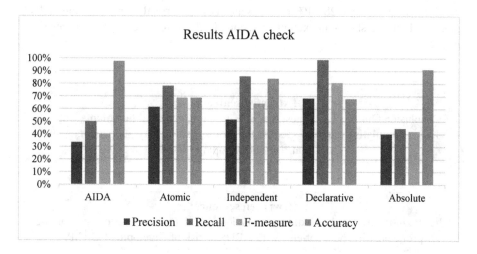

Fig. 2. Results of the AIDA sentence check.

5.3   Rewriting AIDA Sentences

Out of the 87 extracted sentences that contained a claim, not a single one complied with all the requirements of being an AIDA sentence. Table 1 shows the compliancy of the 87 extracted sentences and the rewritten sentences according to our created gold standard.

Table 1. AIDA compliance before and after rewriting.

Requirement	Percentage of compliant sentences before rewriting	Percentage of compliant sentences after rewriting
AIDA	0.00%	28.74%
Atomic	44.83%	48.28%
Independent	12.64%	57.47%
Declarative	74.71%	90.80%
Absolute	2.23%	56.32%

As comes forward from Table 1, the resulting core sentences are improved in many ways after the algorithm has rewritten them. Again, these sentences were manually checked for AIDA compliancy and not by the algorithm to ensure a correct evaluation. Rewriting the sentences shows the best results regarding the independence and absoluteness requirements. Prior to rewriting the sentences, only 2.23% of the sentences was absolute and 12.64% independent. After rewriting the sentences, 56.32% of the sentences comply with the absoluteness requirement and 57.47% with the independence requirement. Overall, 25 out of 87 sentences comply with all the rules after rewriting, meaning that 28.74% of the sentences is correctly rewritten into AIDA sentences. Figure 3 shows an example of a completely correctly extracted and rewritten sentence.

Extracted sentence:
Conclusion Our results suggest that obesity, intraperitoneal fat volume, and a longer cumulative duration spent in the prone position may put patients with ARDS at risk of developing SC-CIP.

Rewritten sentence:
Obesity, intraperitoneal fat volume, and a longer cumulative duration spent in the prone position puts patients with ARDS at risk of developing SC-CIP.

Fig. 3. An example of a completely correctly extracted and rewritten sentence.

6   Discussion

6.1   Extracting Core Sentences

The results of extracting a core sentence from the abstract are promising: in 69.6% of the cases, the algorithm picks a sentence from the abstract that contains a claim. Moreover, in 61.6% it picks the sentence that best describes the core claim. However, when taking into account that an abstract is usually 100–150 words long, it puts the results in perspective. Out of the approximately 10–15 sentences, picking the correct sentence should not be too difficult. For this research, we made the assumption that a scientific article or abstract consists of one core sentence. In the abstract alone, this also often is the case. Sometimes, however, the claim of an article is expressed in two or even more sentences. In some of our 125 random abstracts, multiple sentences showed a claim that was derived from the conducted research. Currently, our algorithm fails to recognize this and picks only the sentence with the highest score. For future improvements, the algorithm could be improved by including the possibility to extract multiple sentences or claims. Building upon this, more recent articles sometimes contain a highlight section that can aid in both the extraction of core sentences but also in the goal of communicating scientific discourse. This section consists of a number of core sentences that show the key findings of the article. Because this section is only present in a minority of the articles, we decided not to look at it but focus on the abstract. In the future, however, this section can be a great help in communicating scientific findings conform the concept of AIDA sentences.

6.2 Checking Sentences for AIDA Compliance

Looking at Fig. 2, the first thing that stands out is the big difference between the accuracy and the other three metrics in the overall AIDA check. This is, in part, due to the fact that out of all extracted sentences, only two comply all the AIDA requirements. Because the algorithm classifies most sentences as not compliant, the accuracy is high. However, when looking at the other three metrics, it shows that the algorithm has a lot of room for improvement. We will discuss the performance of every requirement individually to show performance and potential improvements.

First of all, results of the atomicity check show that all the metrics are relatively equal, with recall being a bit higher than the other three. For this check, the algorithm looks for certain words or phrases and whether the sentence consists of multiple clauses. If it consists of multiple clauses, the sentence can be broken down further and is therefore not atomic. Even though this is done, there are still many cases in which a sentence can be broken down but this is not recognized by the algorithm. Sentence parsing is done with a statistical parser that is not always correct. Due to this, multiple clauses are not always recognized. Improvements for this check can be made with respect to both parsing and the inclusion of more rules.

Second, for the independence check it is evident that precision is lower than the other metrics. With precision being less than the other metrics, it means that relatively many sentences are classified as independent when in fact they are not. To check for independence we only use a list of words and phrases that denote external references. Clearly, this list is not comprehensive enough and to improve this check, this list could be extended but other methods should also be investigated.

Striking from the metrics of the declarativeness check is that the recall is close to 100%, whereas the precision is just below 70%. The current checks for this requirement are minimal because it is hard to check whether a sentence is grammatically correct using a rule-based approach. Because of this, our algorithm classifies almost every sentence as declarative. Therefore, nearly every sentence that truly is declarative, is also classified as such. However, a great deal of sentences that are not declarative are still labeled as declarative – explaining why precision is relatively low. To improve this check, the 'understanding' of sentences should be improved which may prove hard for a rule-based approach like the one we use.

Finally, looking at the metrics of the absoluteness requirement, a similar thing is observed as with the AIDA check as a whole. Out of the 125 extracted sentences, only nine sentences were absolute. Because the algorithm (correctly) labels most of the sentences as not compliant, accuracy is high – over 90%. Precision and recall, however, are lower because the algorithm often fails to recognize that the sentence shows how something was discovered. In these cases, the sentences are classified as compliant when they are not. This check, like others, searches for pre-defined words and patterns that indicate a violation of the requirement. Extending this list could improve the performance of this check.

Overall, the metrics show that most of the checks can be improved in many ways. Including more rules could potentially improve the performance of the algorithm but only up until a certain point. However, our metrics also show the limitations of a rule-based approach. For future improvements, the inclusion and combination with other approaches should be investigated as well.

6.3   Rewriting AIDA Sentences

The results reveal that the algorithm performs rather well when a sentence is not independent or not absolute. These two requirements often go hand in hand. As mentioned before, core sentences often start with an introducing phrase like "*this study suggests that*". This shows both how the results are retrieved (through this study) and is an external reference (to the study). Therefore, these two requirements are closely related. Sentences that are not absolute or not independent are often very well rewritten. As discussed in the Approach, especially when a sentence is not absolute, several things are checked and many changes can be brought about. Probably the most important check (and change) is that of the pattern matching with the regular expression. The introducing phrase mentioned earlier is often present in a core sentence and can therefore in many cases also be removed. For the algorithm to work even better, this pattern matching could be further refined.

The performance of the other two requirements, on the other hand, shows room for improvement. Our algorithm does not even try to deal with sentences that are not atomic and when a sentence is not declarative, some very simple measures are taken. First of all, the algorithm makes sure that the sentence will start with a capital and end with a full stop. Furthermore, headers that are extracted together with the sentence are removed as well to ensure declarativeness. With respect to the atomic requirement, the emphasis was placed on trying to remove clauses from a sentence during training. When sentences are not atomic, it may be because they consist of multiple sub clauses that are irrelevant. However, some taken measures showed to do more harm than good to a lot of sentences and therefore none of the measures were actually implemented. Therefore, when a sentence is not atomic, our current algorithm gives up and does not attempt any rewriting.

7   Conclusion

We presented an attempt to extract core claims from scientific articles. Our algorithm extracts the core sentence of the abstract, checks it for AIDA compliance, and tries to rewrite it in the negative case. In 69.6% of the cases, a sentence is extracted from the abstract that contains a claim from the article. Moreover, in 61.6% of the cases, the perfect core sentence is extracted from the abstract. Evaluating the checks for AIDA compliance was done both individually per requirement and for AIDA as a whole. Because only a few sentences complied with all the AIDA requirements, the accuracy of this check was very high (97.6%) but the F-measure rather low (40.0%). Finally, rewriting a sentence into a fully compliant AIDA sentence was done successfully 28.7% of the time. Overall, these results show that extracting AIDA sentences with a rule-based approach proves difficult but has great potential. With improvements and modifications in the algorithm, the extraction of core claims from scientific articles may become more accurate and efficient in the future. A critical mass of interlinked AIDA sentences made available with text mining might encourage authors to write AIDA sentences for their own findings and to interlink them with claims made by others. This, in turn, will allow us to automatically organize and aggregate scientific findings and discourse, and finally make it easier for researchers to keep up with the latest developments in their scientific field.

References

1. Aggarwal, C.C., Zhai, C. (eds.): Mining text data. Springer Science & Business Media, New York (2012)
2. Barrera, A., Verma, R.: Combining syntax and semantics for automatic extractive single-document summarization. In: Gelbukh, A. (ed.) CICLing 2012, vol. 7182, pp. 366–377. Springer, Heidelberg (2012)
3. Chiticariu, L., Li, Y., Reiss, F.R.: Rule-based information extraction is dead! Long live rule-based information extraction systems! In: EMNLP, pp. 827–832, October 2013
4. Ferreira, R., de Souza Cabral, L., Lins, R.D., e Silva, G.P., Freitas, F., Cavalcanti, G., Lima, R., Simske, S.J., Favaro, L.: Assessing sentence scoring techniques for extractive text summarization. Expert Syst. Appl. **40**(14), 5755–5764 (2013)
5. Hong, B., Zhen, D.: An extended keyword extraction method. Phys. Procedia **24**, 1120–1127 (2012)
6. Kuhn, T.: A survey and classification of controlled natural languages. Comput. Linguist. **40** (1), 121–170 (2014)
7. Kuhn, T., Barbano, P.E., Nagy, M.L., Krauthammer, M.: Broadening the scope of nanopublications. In: Cimiano, P., Corcho, O., Presutti, V., Hollink, L., Rudolph, S. (eds.) ESWC 2013. LNCS, vol. 7882, pp. 487–501. Springer, Heidelberg (2013)
8. Larsen, P.O., Von Ins, M.: The rate of growth in scientific publication and the decline in coverage provided by Science Citation Index. Scientometrics **84**(3), 575–603 (2010)
9. Lloret, E., Romá-Ferri, M.T., Palomar, M.: COMPENDIUM: a text summarization system for generating abstracts of research papers. Data Knowl. Eng. **88**, 164–175 (2013)
10. Mihalcea, R., Tarau, P.: TextRank: Bringing Order into Texts. Association for Computational Linguistics, Barcelona (2004)
11. Mons, B., van Haagen, H., Chichester, C., den Dunnen, J.T., et al.: The value of data. Nat. Genet. **43**(4), 281–283 (2011)
12. Ramos, J.: Using tf-idf to determine word relevance in document queries. In: Proceedings of the First Instructional Conference on Machine Learning, December 2003
13. Rose, S., Engel, D., Cramer, N., Cowley, W.: Automatic keyword extraction from individual documents. Text Mining, pp. 1–20 (2010)
14. Saggion, H., Poibeau, T.: Automatic text summarization: past, present and future. In: Poibeau, T., Saggion, H., Piskorski, J., Yangarber, R. (eds.) Multi-Source, Multilingual Information Extraction and Summarization, pp. 3–21. Springer, Heidelberg (2013)
15. Shah, P.K., Perez-Iratxeta, C., Bork, P., Andrade, M.A.: Information extraction from full text scientific articles: Where are the keywords? BMC Bioinform. **4**(1), 20 (2003)

16. Siddiqi, S., Sharan, A.: Keyword and keyphrase extraction techniques: a literature review. J. Comput. Appl. **109**(2), 18–23 (2015)
17. Tan, A.H.: Text mining: the state of the art and the challenges. In: Proceedings of the PAKDD 1999 Workshop on Knowledge Discovery from Advanced Databases 8, pp. 65–70 (1999)
18. Turney, P.D.: Learning algorithms for keyphrase extraction. Inform. Retrieval **2**(4), 303–336 (2000)
19. De Waard, A., Schneider, J.: Formalising uncertainty: an ontology of reasoning, certainty and attribution (ORCA). In: Proceedings of the Joint 2012 International Conference on Semantic Technologies Applied to Biomedical Informatics and Individualized Medicine, vol. 930, pp. 10–17. CEUR-WS.org., November 2012
20. Wartena, C., Brussee, R., Slakhorst, W.: Keyword extraction using word co-occurrence. In: 2010 Workshop on Database and Expert Systems Applications (DEXA), pp. 54–58. IEEE, August 2010
21. Zweigenbaum, P., Demner-Fushman, D., Yu, H., Cohen, K.B.: Frontiers of biomedical text mining: current progress. Brief. Bioinform. **8**(5), 358–375 (2007)

Towards Legal Compliance by Correlating Standards and Laws with a Semi-automated Methodology

Cesare Bartolini[1]([✉]), Andra Giurgiu[1], Gabriele Lenzini[1], and Livio Robaldo[1,2]

[1] Interdisciplinary Centre for Security, Reliability and Trust (SnT),
University of Luxembourg, Luxembourg, Luxembourg
{cesare.bartolini,andra.giurgiu,gabriele.lenzini,livio.robaldo}@uni.lu
[2] Computer Science and Communications Research Unit (CSC),
University of Luxembourg, Luxembourg, Luxembourg

Abstract. Since generally legal regulations do not provide clear parameters to determine when their requirements are met, achieving legal compliance is not trivial. The adoption of standards could help create an argument of compliance in favour of the implementing party, provided there is a clear correspondence between the provisions of a specific standard and the regulation's requirements. However, identifying such correspondences is a complex process which is complicated further by the fact that the established correlations may be overridden in time *e.g.,* because newer court decisions change the interpretation of certain legal provisions. To help solve these problems, we present a framework that supports legal experts in recognizing correlations between provisions in a standard and requirements in a given law. The framework relies on state-of-the-art Natural Language Semantics techniques to process the linguistic terms of the two documents, and maintains a knowledge base of the logic representations of the terms, together with their defeasible correlations, both formal and substantive. An application of the framework is shown by comparing a provision of the European General Data Protection Regulation with the ISO/IEC 27018:2014 standard.

Keywords: Legal compliance · Legal requirements · Security standards · General data protection regulation

1 Introduction

As it happens with the European Union (EU) harmonized standards used to demonstrate that products, services, or processes comply with relevant EU legislation [11], when a standard published by a standardization body is endorsed by the law, then implementing the standard also gives a *legal presumption of compliance*. However, harmonized standards are a fortunate but uncommon case. More often, standards do not have such a direct effect on legal compliance. By adopting a standard however, an organisation can demonstrate a proactive attitude and best efforts to be compliant according to the state of the art in that

© Springer International Publishing AG 2017
T. Bosse and B. Bredeweg (Eds.): BNAIC 2016, CCIS 765, pp. 47–62, 2017.
DOI: 10.1007/978-3-319-67468-1_4

specific domain. Standards can thus provide the organisation with an argument of compliance.

Such an argument of compliance, would rely on proving a clear correspondence between the provisions of a specific standard and the law's requirements. But identifying such correspondences is not easy. It is also a dynamic process where the established correlations can be further overridden in time, for instance because newer court decisions change the interpretation of certain legal provisions.

In this paper, we propose a way to ease, and partly to automate, the process of checking for such document-to-document correspondences. We discuss a software framework that aids in determining the formal and substantive correlations between the provisions in a standard and those in a law. The framework's core is a logic-based methodology to represent, in a machine-processable format, (a) the relevant syntactic concepts in the provisions, and (b) the relevant correlations between them. In this paper, we describe this logic-based methodology and exemplify how it works using provisions from two specific and relevant documents. One is the standard ISO/IEC 27018:2014 (ISO 27018, in short), which concerns public clouds acting as personal data processors. This standard can be regarded as a building block [11] that helps data-processing organizations comply with the principle of data protection accountability. The second document is the General Data Protection Regulation (GDPR), which is the new law on data protection in the EU (see Sect. 2).

The framework depends on two auxiliary functional blocks (see Sect. 4): *i* a *logic knowledge base*, which can be populated, corrected and extended by legal experts and that stores the machine-processable logic correlations; *ii* a *set of Natural Language Semantics (NLS) and Natural Language Processing (NLP) techniques*, which allow a user to browse a XML representation of the documents and to search and retrieve the words, the terms, and the sentences that have been found relevant for correlation. The NLS and NLP techniques help users to efficiently and precisely find the established correlations within the knowledge base, which any expert user can successively reinforce, correct, justify, and expand. The selection of relevant terms and the definition of correlations, requiring human reading, processing and decision-making, is therefore semi-automatic.

The correlations are expressed formally in a deontic and defeasible logic for legal semantics called *Reified Input/Output Logic* (see Sect. 3). Defeasibility is required since, due to differences in legal interpretation, some of the correlations could be in contradiction: interpretations from more authoritative sources (such as high courts) are thus required to eventually resolve the conflict.

2   The Data Protection Reform

The GDPR, which will apply from 25 May 2018, replaces the current Directive 95/46/EC[1] with more modern rules, better adapted to the data processing real-

---

[1] Directive 95/46/EC of the European Parliament and of the Council of 24 October 1995 on the protection of individuals with regard to the processing of personal data and on the free movement of such data.

ities of today. Unlike Directive 95/46/EC, which required implementation by the Member States, the GDPR will be directly applicable and does not require transposition into national law.

The purposes of the GDPR [30] are to align data protection rules with the most recent developments in data-processing technologies while still providing a legislation that is flexible enough not to become outdated over the course of a few years. The Regulation enhances the responsibility of data controllers and strengthens the rights of the data subject. Controllers will face heavy administrative fines in case of non-compliance with its provisions [13], which go as high as four percent of the total worldwide annual turnover of an undertaking[2].

As many of the provisions of the GDPR are quite general and potentially applicable in diverse ways, its interpretation through legal doctrine and jurisprudence will be of essence. The Regulation won't be applicable before 25 May 2018 and therefore no decisions based on its provisions can exist until then. When it will be applied, relevant decisions on its interpretations are expected to be issued by the Data Protection Authorities (DPAs) of Member States, by national courts, and by the Court of Justice of the European Union (CJEU).

3   Related Work

The application of NLP and NLS to the legal domain is a research trend that has received a lot of attention and investments in recent years, as shown by several acU-funded projects on the topic such as Openlaws[3], ProLeMAS[4], and MIREL[5]. Modern NLP technologies [6] are able to classify and discover inter-links between legal documents thanks to parsers [1], statistical algorithms [8], and legal terminological databases or legal ontologies [8,40]. This is often done by transforming the legal documents into XML standards, such as *Akoma Ntoso*[6], where relevant information are tagged. An example of commercial legal document management system employing these technologies is *Eurocases*[7], which collects EU case law and uses NLP techniques to classify the documents on the basis of their topic [7].

Although these systems help navigate legislative documents and retrieve information, their overall usefulness is limited because they process words disregarding their possible different semantic interpretations. The latter would allow for legal reasoning, *e.g.*, correlating laws among them and determining whether they lead to inconsistencies. Semantic processing of documents like laws/regulations and security standards is what we are going to propose as a component of the methodology we present in Sect. 4.

The underlying logical framework we will use in our methodology is reified Input/Output logic [36], a recent approach designed as an attempt to investigate the *logical architecture* of the provisions in natural language. Reified

---

[2] GDPR, Article 83.
[3] https://info.openlaws.com/openlaws-eu/.
[4] http://www.liviorobaldo.com/Prolemas.htm.
[5] http://www.mirelproject.eu/.
[6] http://www.akomantoso.org/.
[7] http://eurocases.eu/.

Input/Output logic merges Input/Output logic [24], a well-known formalism in Deontic Logic (*i.e.*, a logic that expresses concepts like permissions, obligations, prohibitions), with the First Order Logic (FOL) for NLS proposed by [18], which is grounded on the concept of *reification*.

3.1 Reification and Input/Output Logic

Reification [10] is a concept that allows to move from standard notation in FOL such as "(give a b c)", asserting that "a" gives "b" to "c", to another notation in FOL "(give$'$ e a b c)", where "e" is the *reification* of the giving action. "e" is a FOL term denoting the giving event of "b" by "a" to "c". Reification is able to express a wide range of phenomena in NLS via simple *flat* FOL formulæ, which are basically conjunctions of atomic first-order predicates.

It has been argued [19,31,32] that flat logical formulæ for NLS have a twofold advantage. First, they allow to properly represent the semantics of several linguistic constructions which are hard to represent in other popular formalisms for NLS, such as Discourse Representation Theory (DRT) [23] or Minimal Recursion Semantics (MRS) [9]. Those formalisms introduce complex operators, *e.g.*, modal or causal operators, which take subformulæ as an argument. Nesting sub-formulæ within other (sub-)formulæ prevents several readings indeed available in NLS, such as cumulative readings (see [33]) or causality and concession (see [17,35]).

Secondly, flat formulæ enhance human readability and comprehension as well as the ease of controlling computational complexity. These two features are of course pivotal in the construction and debugging of large knowledge bases of formulæ, to be used in practical applications, as advocated in our methodology.

Let us consider a simple example: the representation of the sentence "Jack wants to eat an ice cream". In standard NLS formalisms, *e.g.*, DRT and MRS, this sentence is formalized by introducing a modal operator for representing the verb "want", *e.g.*, "want(... , ...)", which applies to a sub-formula representing the sentence "Exists an ice cream that Jack eats". For instance:

$$\mathsf{want}(\mathsf{Jack}, \exists_{ic}(\mathsf{eat}(\mathsf{Jack}, \mathsf{ic})))$$

In Hobbs's, the eating hypothetical action is reified into an eventuality that is inserted as argument of a FOL predicate *want$'$*:

$$\exists_e \exists_{e1} \exists_{ic}[\,(\mathsf{Rexist}\ e) \land (\mathsf{want}'\ e\ \mathsf{Jack}\ e_1) \land (\mathsf{eat}'\ e_1\ \mathsf{Jack}\ ic)\,]$$

Rexist is a special predicate used to assert which eventualities really exist in the context; in the example above, the wanting event really exists while the eating event does not. In the formulæ below, we will omit the Rexist predicates for space constraints; we will deem easy to understand, in those formulæ, which eventualities really exist and which do not.

On the other hand, Input/Output logic [24] is a well-known formalism in deontic logic, grounded on norm-based semantics. Norm-based semantics has been proposed as an alternative to deontic frameworks based on possible-world semantics, such as STIT logic [20] and dynamic deontic logic [27]. It has been argued that norm-based semantics provides: (1) a more flexible handling of the well-known Jørgensen's dilemma [22], stating that, contrary to declarative statements, norms do not corresponds to truth values, *i.e.*, they cannot be described as true or false; (2) a straightforward and simple way to deal with moral conflicts and different kinds of permissions (see [25,29] to see how to deal with these in input/output logic); (3) a simpler and easy-to-control complexity; it has been argued [39] that compliance checking in Input/Output logic are coNP/NP hard and in the second level of the polynomial hierarchy, while STIT and dynamic deontic logic are respectively undecidable and EXPTIME-complete.

Input/output logic is not the only deontic framework that is not based on possible-world semantics; alternatives are imperative logic [16], prioritized default logic [21] and defeasible deontic logic [14]. A fine-grained comparison between the mentioned formalisms on the basis of norm-based semantics goes beyond the scope of this paper, in that it mostly lies on a theoretical level. Therefore, we address the interested reader to the relevant literature.

Input/output normative systems are triples $N = (O, P, C)$ of three sets of pairs: obligations (O), permissions (P), and constituency rules (C). A pair (a, b) corresponds to an if-then rule. The expression $(a, b) \in O$ reads "if a holds, then b is obligatory"; $(a, b) \in P$ reads "if a holds, then b is permitted"; and $(a, b) \in C$ corresponds to the standard FOL implication "$a \rightarrow b$". For space constraints, below we will not consider permissions (see [25]).

So far, Input/Output logic has been mostly studied from a theoretical point of view, with the elements a and b of the pairs being formulæ in propositional logic. Reified Input/Output logic is the first attempt to make Input/Output logic usable in practical applications in legal informatics, and to be used for representing norms from existing legislation available in natural language only.

3.2 Reified Input/Output Logic

As said above, reified Input/Output logic combines the advantages of the reified and of the Input/Output logic, first of all their respective formal simplicity. As argued in [36], simplicity appears to be a necessary feature for a logical formalism designed for application in legal informatics, in order to foster active collaboration of legal practitioners, usually having little expertise in logic, who can contribute to the building of large knowledge bases of formulæ. The methodology illustrated here represent the first attempt to build such a knowledge base, with respect to the data protection domain.

In reified Input/Output logic, a and b are formulæ as defined in [18]. Thus a sentence like "every bird who wants to eat is obliged to tweet" is represented as:

$$\forall_x (\exists_e \exists_{e_1} [(\text{bird } x) \wedge (\text{want}' \ e \ x \ e_1) \wedge (\text{eat}' \ e_1 \ x)], \ \exists_{e_2} [(\text{tweet}' \ e_2 \ x)]) \in O$$

Obligations are intended to populate the ABox of the knowledge base, *i.e.*, the set of assertive contextual statements. On the other hand, the set of definitions, axioms, and constraints on the predicates used in the ABox are part of the TBox of the knowledge base, *i.e.*, the set of terminological declarative statements, also known as constitutive rules.

As said above, constitutive rules are expressed as standard FOL implications. For the sake of readability, below we will assert them as such (*i.e.*, in the form "$a \rightarrow b$"), rather than as pairs in the form (a, b). For instance, the TBox could contain a FOL implication specifying that all birds fly:

$$\forall_x (\,(\text{bird } x) \rightarrow \exists_e[\,(\text{fly}' \; e \; x)\,]\,) \in C$$

Finally, following [18], in reified Input/Output logic the implications populating the TBox can be made defeasible via a mechanism drawn from Circumscriptive Logic [26]. The antecedent of the implications can include an extra predicate that can be *assumed* or not. For instance, the previous implication can be made defeasible by adding a predicate "*assumption$_1$*":

$$\forall_x (\,(\,(\text{bird } x) \wedge (\text{assumption}_1 \; x)\,) \rightarrow \exists_e[\,(\text{fly}' \; e \; x)\,]\,) \in C$$

This formula reads as: "if x is a bird and it can be assumed that x is a 'normal' bird, then x flies". Not for all birds the assumption can be made. For instance, penguins are a special type of birds that do not fly. This is codified as:

$$\forall_x (\,(\text{penguin } x) \rightarrow (\,(\text{bird } x) \wedge \neg(\text{assumption}_1 \; x)\,)\,) \in C$$

If x is penguin, we derive that x is a bird but we cannot derive that x flies.

In our knowledge base, assumptions are used to model legal interpretations. Handling legal interpretations via defeasible mechanisms is a common practice in logical frameworks for legal informatics, such as [15]. A separate part of our knowledge base will specify which legal authorities adopt a certain assumption, which ones do not, and which ones either adopt it or not under certain conditions. By selecting certain legal interpretations, it is then possible to derive different conclusions from the knowledge base.

4   Methodology

The framework we propose offers a computer-aided methodology to analyze standards to make an argument of compliance with respect to a specific piece of legal text, and is schematically summarized in Fig. 1. Users (who may be lawyers, regulators, auditors, or other legal experts) access a digital and annotated XML representation of the normative texts (laws and standards). While browsing a document and selecting the relevant concepts, NLP and NLS tools help traverse the rest of the documents, find related terms, and recall previous correlations between them. Correlations from different sources have different degrees of importance, which need to be tracked using specific metadata. The framework implements a collaborative strategy to evaluate the stored correlations. The user's decisions are stored in the knowledge base, after being appropriately represented in a logic for legal semantics.

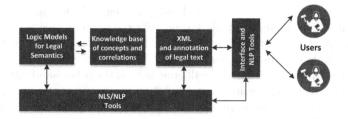

Fig. 1. The framework at a glance.

The framework, and in particular its knowledge base, does not pretend to be complete. It rather provides expert users with an updated knowledge that helps take autonomous and informed decisions, both when confirming the correlations the tool suggests and when choosing to define new correlations. The knowledge base is designed to support defeasible reasoning, *i.e.*, to tolerate (apparent) inconsistencies of different interpretations of terms, by overriding general assertions into more contextually-specific ones. Conflicts are especially frequent in legal interpretation, but they can generally be solved considering that the interpretation by higher-instance courts, such as the CJEU, prevails over lower ones. In order to cope with interpretations of different legal weights, which may supersede one another, the logic formalism that the framework embeds is defeasible: correlations can be updated, modified, rewritten and weighed. If conflicts do remain, the framework still embeds strategies that help the user take a decision.

As pointed out in Sect. 1, correlations can be further divided into two different categories: *formal* and *substantive* correlations.

Formal correlations entail a mere textual overlap between concepts. For example, formal correlations would allow us to observe that both the GDPR and the ISO 27018 standard use the term "notify" (see Sect. 5). Substantive correlations are more complex and entail the analysis of the actual meaning of terms. To assert a correlation of this kind, requirements must be met in a concrete way. Following the previous example, to assert a correlation between a provision of the GDPR and one of the standard concerning notification, it is necessary to verify the exact meaning of the term "notify" in the two texts.

The methodology follows three steps to build the correlations, involving both a legal and a technical approach. The legal approach is focused on the interpretation of the provisions of laws (the GDPR in our example) and standards (ISO 27018), whereas the technical approach consists of modelling those provisions into an ontology, and expressing the interpretation by means of logical formulæ.

Step 1: analysis of the provisions. The provisions of the law and of the security standard are analyzed by legal experts, who provide an interpretation of the terms used in the provisions and compare them in search of semantic correlations. There is no need for this interpretation to be final, as more interpretations can be added later, and old interpretations can be overridden by newer ones, but this start requires a significant manual activity.

This step entails two sub-steps. First, the legal documents need to be expressed in a machine-processable format. For the scope of this work, we have selected the `Akoma Ntoso` language[8]. This is an XML-based format that allows to easily navigate the documents and identify the relevant provisions in the legal text. The legal interpretations of the documents, on the other hand, are stored in separate documents, and contain a reference to their source, and another reference to the provision, or set of provisions, to which they apply. The use of unique namespaces and identifiers in `Akoma Ntoso` allows for a fine-grained model of a legal text and its interpretations.

To support the execution of this step, legal experts are assisted by external NLP procedures that suggest (semi-automatically and during the browsing of the documents) previous translations and correlations on the basis of the information currently stored in the knowledge base. Ultimately, it is the legal expert who must decide whether and how correlations need to be overridden. In that case, new correlations are added and annotated with the source which contributed to define it (*e.g.,* Court of Justice of the European Union, *Dapreco and Copreda Corp.*, C-XYZ/16). Implications by more authoritative sources override defeasible implications by less authoritative ones.

The interaction between the legal expert and the knowledge base will have to take into account that the former are not expected to have deep technical knowledge. This will be easily managed by means of appropriate front-ends and user interfaces that can be serve the purpose of viewing and modifying existing interpretations and their connections with specific provisions, and that of adding new interpretations, specifying metadata such as their nature, source, and validity in space and time.

Step 2: creation of legal ontologies. The legal interpretations are mapped onto legal ontologies of the law and the security standard. Legal ontologies [5] model the legal concepts, parties and stakeholders affected by the law, the duties and rights of each stakeholder, and the sanctions for violating the duties. As per ontologies in general, legal ontologies are expressed in a knowledge representation language. For this work, we chose the popular abstract language OWL, which can be serialized using various XML notations. For example, the OWL representation of the data protection ontology will contain concepts such as "controller", "data subject", "personal data", "processing" and so on.

The ontologies operate as the semantic base for the formal representation of the legal documents and their interpretation expressed in the form of logic formulæ. In other words, the ontologies represent the pivot of the methodology, as both normative documents and the objects of the logic (predicates and terms) are connected to the concepts expressed in the ontologies. In our final knowledge base, the connection will be implemented via `LegalRuleML` OASIS standard[9], along the same lines illustrated in [12].

---

[8] http://www.akomantoso.org/.

[9] https://www.oasis-open.org/committees/legalruleml.

The use of the OWL language guarantees that every concept in the ontologies is uniquely identified, thus creating an unambiguous connection between a logic formula, on the one side, and the concept or concepts to which it relates, on the other. Using this connection, it will be easy to perform searches based not solely on textual content, but also on semantic concepts and the relations between them. Consequently, a search for legal interpretations can overcome linguistic barriers such as typos or synonyms, and even be language-independent.

Work has started towards the creation of an improved ontology. In the interim, a preliminary version of a legal ontology for the GDPR has been defined already [4]. Albeit partial and based on an older version of the GDPR, it was designed to express the duties of the controller. As such, it can be used to find the correspondences between the requirements expressed in the GDPR and in security standards.

Step 3: generation of logic formulæ. The third and final step of the methodology consists of generating the logical formulæ representing the set of provisions in the law and the set of provisions in the security standard, as well as the implications between them. These formulæ are expressed in reified Input/Output logic [34]. An example is shown in the next section.

Associating textual provisions to logical formulæ amounts to converting ambiguous and vague terms into non-ambiguous items (predicates and terms). Words in the provisions are represented via predicates reflecting their vagueness. For example, the word "notify", included in the sample provisions used in Sect. 5, will be represented via the homonym predicate "notify". These "vague" predicates may be defined by adding implications and further constraints (axioms). Those implications will be *defeasible*, so that they can account for different legal interpretations. Predicates are associated with classes of the ontologies developed in Step 2 or with standard general-purpose ontologies/repositories belonging to the NLS literature, *e.g.*, Verbnet [38].

5   Generation of Logic Formulæ: Example

We exemplify step 3 of our methodology. This step, which lies at the core of the methodology, is the most technical, and more innovative than steps 1 and 2 which instead rely upon existing techniques. We use a provision from the GDPR and an article of the ISO 27018 security standard:

(*a*) GDPR, Article 33.2: *The processor shall notify the controller without undue delay after becoming aware of a personal data breach.*
(*b*) ISO 27018, Article A9.1: *The public cloud PII processor should promptly notify the relevant cloud service customer in the event of any unauthorized access to PII.*

The formulæ will include predicates reflecting the vagueness of the terms occurring in the sentences. Thus, for instance, the verb "notify", which occurs in both provisions, is formalized into an homonymous predicate "notify".

On the other hand, the provisions in the ISO 27018 use the term "PII" (Personally Identifiable Information) while the ones in the GDPR use the term "personal data". Although one might simply consider the two terms as synonyms (as suggested in ISO 27018, Article 0.1), and thus associate them with the same predicate, our methodology keeps them distinct, *i.e.*, it formalizes them via two different predicates "personalData" and "PII". An additional axiom is then added to the TBox, in order to correlate the two predicates:

$$\forall_x \left[(\text{PII } x) \rightarrow (\text{personalData } x) \right] \tag{1}$$

The implication can be made defeasible by adding an assumption as shown in Sect. 3. In that case, the normal rule is that PII is also considered personal data, unless there is a special exception which overrides the general rule.

In light of this, the GDPR provision in (*a*) is formalized as follows:

$$\forall_{e_b} \forall_x \forall_y ($$
$$\exists_{e_p} \exists_{e_c} \exists_{e_a} \exists_z \left[(\text{dataProcessor } x) \land (\text{dataController } y) \land (\text{personalData } z) \land \right.$$
$$(\text{process}' \ e_p \ x \ z) \land (\text{control}' \ e_c \ y \ z) \land (\text{awareOf}' \ e_a \ x \ e_b) \land (\text{dataBreach } e_b \ z)],$$
$$\left. \exists_{e_n} \left[(\text{notify}' \ e_n \ x \ y \ e_b) \land (\text{nonDelayed } e_{nd} \ e_n) \right] \right) \tag{2}$$

In (2), "e_p", "e_c", "e_a", 'e_b", and "e_n" are variables referring to events. "e_p" is the event of processing the personal data "z" (patient) performed by the data processor "x" (agent). "e_c" is the event[10] of controlling the personal data "z" (patient) performed by the data controller "y" (agent). If x become aware ("e_a") of an event of data breach ("e_b") of the personal data "z", then x is obliged to notify it ("e_n") to the data controller and that event must be done with undue delay (predicate "nonDelayed", applied to the eventuality "e_n").

In (2), "notify" and "nonDelayed" are predicates whose meaning is subject to different legal interpretations. Recalling the difference between *formal* and *substantive* compliance outlined in Sect. 1, we note that the formalization in (2) only enforces formal compliance. The formula in (2) simply requires the data processor to notify data breaches without undue delay, but it does not specify *how* notifications should be performed for being legitimate.

For instance, the data processor could require the data controller to acknowledge the notification, in order to make sure it was received. Similarly, the processor could be required to avoid sending notifications of data breaches via standard paper mail, in that the time needed by the postal service to deliver the mail could be considered as an undue delay. It is up to judicial authorities to establish the substantive compliance of the obligation in (2).

Of course, we do not have the authority to decide whether (2) is performed in the proper way. In our work, we only aim at providing a methodology to keep track of all legal interpretations of the provisions. From a formal point of view,

---

[10] In this context, an event must not be considered as a specific occurrence happening at a given time, but as a wider concept encompassing the whole of the controller's activity.

the TBox must be enriched with axioms defining the conditions under which the predicates in the formula are true. Those axioms are defeasible, therefore they will contain *assumptions* that may be taken or not. And, it is possible to separately assert that certain assumptions are taken to be either true or false by certain legal authorities, possibly under certain further conditions (see below).

For instance, by assuming that email with electronic signature is a proper and prompt means to notify the data controller, the following (defeasible) axiom is added to the TBOX.

$$\forall_x \forall_y \forall_{e_1} \forall_{e_2} [((\mathsf{sendEmailWithES}\ e_1\ x\ y\ e_2) \wedge (\mathsf{assumption}_2\ e_1)) \rightarrow$$
$$\exists_{e_n} ((\mathsf{notify}'\ e_n\ x\ y\ e_2) \wedge (\mathsf{nonDelayed}'\ e_{nd}\ e_n)) \tag{3}$$

Formula (4) models the ISO 27018 provision in (*b*):

$$\forall_e \forall_x \forall_y \forall_z \forall_{e_p} \forall_{e_c} \forall_{e_b} (((\mathsf{PIIProcessor}\ x) \wedge (\mathsf{PIIController}\ y) \wedge (\mathsf{PII}\ z) \wedge$$
$$(\mathsf{process}'\ e_p\ x\ z) \wedge (\mathsf{control}'\ e_c\ y\ z) \wedge (\mathsf{access}'\ e_a\ z)) \wedge (\mathsf{unauthorized}\ e_a)),$$
$$\exists_{e_n} [(\mathsf{notify}'\ e_n\ x\ y\ e_a) \wedge (\mathsf{promptly}'\ e_{np}\ e_n)]) \tag{4}$$

As it was done for formalizing (*a*) into (2), the formula introduces predicates that reflect the generic terms used in the text. With an important exception: we formalized "the relevant cloud service customer" via the predicate "PIIController". According to ISO 27018 the cloud service customer can be either a natural person, "PII principal" or a "PII controller", which processes the PII relating to PII principals.[11]

Therefore, building our example on the specific provision of the GDPR, the cloud service provider ("PII Processor" - "data processor") handling data of natural persons ("PII Principals" - "data subjects") will have to notify the organization on behalf of which it processes the data ("PII controller" - "data controller") of occurring incidents ("any unauthorized access to PII" - "personal data breach") in a specific time frame ("without undue delay after becoming aware" - "promptly").

The final ingredient needed for (4) and (2) are axioms relating to the predicates occurring in both, similar to the axiom in (1), which state that PII is by default considered as personal data:

$$\forall_x [(\mathsf{PIIProcessor}\ x) \rightarrow (\mathsf{dataProcessor}\ x)],$$
$$\forall_x [(\mathsf{promptly}'\ x\ y) \rightarrow (\mathsf{nonDelayed}'\ x\ y)],$$
$$\forall_x [(\mathsf{PIIController}\ x) \rightarrow (\mathsf{dataController}\ x)],$$
$$\forall_e \forall_z [((\mathsf{access}'\ e\ z) \wedge (\mathsf{unauthorized}\ e)) \rightarrow (\mathsf{dataBreach}\ e\ z)] \tag{5}$$

The axioms in (5) are quite intuitive. For instance, the first one states that any entity that is considered a PII processor according to ISO 27018 is also a data processor with respect to the GDPR.

---

[11] ISO 27018, Article 0.1: "The cloud service customer, who has the contractual relationship with the public cloud PII processor, can range from a natural person, a 'PII principal', processing his or her own PII in the cloud, to an organization, a 'PII controller', processing PII relating to many PII principals".

It is easy to verify that every tuple of variables "e", "x", "y", "z", "e_p", "e_c", and "e_a" that satisfies formula (4) also satisfies formula (2).

Again, the correlations in (5) can be made defeasible, in order to encompass different legal interpretations of the provisions. For instance, the fact that an unauthorized access *is* a data breach depends on the legal interpretation (which can vary according to the court or authority). According to Article 4(12) of the GDPR, which is essentially equivalent to Article 3.1 of ISO 27018, a data breach is defined as follows:

GDPR, Article 4(12): *'personal data breach' means a breach of security leading to the accidental or unlawful destruction, loss, alteration, unauthorized disclosure of, or access to personal data transmitted, stored or otherwise processed.*

To encompass the different legal interpretations of Article 4(12), we enrich the last implication in (5) via a predicate stating that the eventuality e is both an unauthorized access and a data breach under general conditions, but there might be exceptions where it is an unauthorized access, but not a data breach:

$$\forall_e \forall_z [\, (\, (\text{access}'\ e\ z) \wedge (\text{unauthorized } e) \wedge (\text{assumption}_e\ e)\,) $$
$$\rightarrow (\text{dataBreach } e\ z)\,] \tag{6}$$

Then, in the knowledge base we separately assert that there are several possible parallel interpretations concerning the assumption. In particular, a court or authority might decide that:

- the assumption holds as true;
- the assumption does not hold;
- the assumption can hold or not, depending on the conditions.

As an example, we can assume fictitious case law, *not* pertaining to actual legal decisions, for the sole purpose of illustrating the methodology. We make up three decisions as follows:

- Italian *Corte di Cassazione, sezione civile, 12530/2012;*
- Spanish *Audiencia provincial de Toledo, n. 57/2016, 2/12/2016;*
- French *Tribunal de Grande Instance d'Avignon, décision du 17/04/2016.* In this case, we assume that the specific conditions examined by the Tribunal consisted in the company *Alpha* performing a security test on an IT system; even if unauthorized accesses indeed took place, those cannot be taken as data breaches in that they were part of the security test.

Table 1 displays the interpretations by the three sources, and the way they are represented by means of a formula in Reified Input/Output logic. Depending on the legal interpretation of Article 4(12) that is selected, different inferences are enabled on the knowledge base.

LegalRuleML provides tags to represent different legal interpretations of the logical items (see [3]). In our future work, we plan to enrich the knowledge base in LegalRuleML with legal interpretations, as soon as they come available.

Table 1. Samples of legal interpretations and their translations.

Source	Interpretation	Formula
Cassazione civile	An unauthorized access is a data breach	$(\mathsf{assumption}_e\ e)$
Audiencia de Toledo	A data breach requires not only an unauthorized access, but also a breach of security and a causal connection between them	$\neg(\mathsf{assumption}_e\ e)$
Tribunal d'Avignon	When a breach of security is part of security tests, such as the ones performed by the company *Alpha*, leading to unauthorized access, it is not considered as a data breach	$\forall_e(\exists_z\exists_{e_t}[\,((\mathsf{access}\ e\ z)\ \wedge$ $(\mathsf{unauthorized}\ e)\ \wedge$ $(\mathsf{partOf}\ e\ e_t)\ \wedge$ $\mathsf{securityTest}\ e_t))\,]\rightarrow$ $\neg(\mathsf{assumption}_e\ e)\,)$

6 Discussion and Conclusion

This paper deals with the complex problematic of complying with abstract legal rules that usually give little guidance as to the implementation of technical measures to address the requirements therein. At the same time, it intends to show the benefits of standards as regards the argument of compliance they can create in favour on the implementing party once a bridge between the law and the standards has been created.

This paper advances a logic formalism whereby correlations between provisions of the law and those of a standard can be expressed, and then introduces a methodology to help build such a bridge in a semi-automated way. The paper exemplifies the methodology in the context of data protection laws (in particular, the GDPR) and security standards (ISO 27018), two domains that significantly overlap given that security is an inherent part of data protection.

By following the illustrated methodology, one can build a machine-processable *knowledge base* of logic formulæ that model and store relevant concepts from a law and a standard together with their possible formal correlations. The knowledge base, which will be collaboratively accessible, will have its records updated and labelled considering the outcomes of specific auditing processes or decision of the courts; in time, it will embed substantive correlations, which can serve as a base for a legal argument for compliance. The logic into which we translate the correlations is defeasible, allowing them be overridden.

The methodology herein is currently a work in progress and not fully implemented yet: this will require the definition of a detailed taxonomy of concepts extracted from the law and the security standard. As the work presented herein is part of a larger research project, its extension is envisioned along various research directions.

Several technical challenges related to building and updating the knowledge base are raised. The translations from natural language to logical formulæ must be uniform for similar text excerpts. To achieve this, we must overcome the limitations of a manual translation, which would be time-consuming and error-prone. For this reason, our work must rely on NLP technologies. However, even at the best of their performances, current NLP algorithms are still unable to automatically carry out the translation with a reasonable level of accuracy, so we advocate a *semi-automatic* translation of the provisions. Similar approaches are applied to translations in general, where translators are helped by collaborative tools such as the "SDL Trados Studio"[12], which suggests, via text-similarity or pattern-matching NLP techniques (see [28,37]), how to translate a sentence on the basis of the translations of similar sentences that the translators have previously stored in the tool. Inspired by that approach, we will develop an enhanced text editor to assist the manual translation of provisions into formulæ. For each provision, the editor will display the translations of similar provisions found via NLP procedures applied to the provisions already stored in the knowledge base, in order to induce uniform translations for similar provisions.

In its current structure, the knowledge base does not rely on the connection between the legal interpretations and the ontology of concepts. Without such a connection, the knowledge base is not yet mature to retrieve legal interpretations. To achieve this goal an additional layer will be needed, some formalism that creates a link between the interpretations and their related concepts. A potential candidate for this purpose has been identified in the XML language LegalRuleML [2].

Additionally, we are aware that the knowledge base must be consistent, *i.e.,* without contradictions, even after having applied the defeasibility measures. To check for consistency, we plan to store formulæ using a XML-based data model, and employ/extend reasoners to monitor the consistency of the knowledge base, whenever new formulæ are added to it.

Finally, for a solid population of the knowledge base, the methodology would greatly benefit from a close interaction with legal authorities. We are envisioning such an interaction in the near future.

Acknowledgments. This work is financed by the Luxembourg National Research Fund (FNR) CORE project C16/IS/11333956 "DAPRECO: DAta Protection REgulation COmpliance". Robaldo has received funding from the EU Horizon 2020 Programme for Research and Innovation under the Marie Skłodowska-Curie grant agreement No. 690974 for the project "MIREL: MIning and REasoning with Legal texts".

References

1. Arora, C., Sabetzadeh, M., Briand, L.C., Zimmer, F.: Automated checking of conformance to requirements templates using natural language processing. IEEE Trans. Software Eng. **41**(10), 944–968 (2015)
2. Athan, T., Boley, H., Governatori, G., Palmirani, M., Paschke, A., Wyner, A.: OASIS LegalRuleML. In: Proceedings of the Fourteenth International Conference on Artificial Intelligence and Law (ICAIL), pp. 3–12. Association for Computing Machinery (ACM), June 2013

---

[12] http://www.translationzone.com/products/trados-studio/.

3. Athan, T., Governatori, G., Palmirani, M., Paschke, A., Wyner, A.: LegalRuleML: design principles and foundations. In: Faber, W., Paschke, A. (eds.) Reasoning Web 2015. LNCS, vol. 9203, pp. 151–188. Springer, Cham (2015). doi:10.1007/978-3-319-21768-0_6

4. Bartolini, C., Muthuri, R., Santos, C.: Using ontologies to model data protection requirements in workflows. In: Proceedings of the 9th International Working on Juris-informatics (JURISIN). pp. 27–40, extended version to be published in LNAI book, November 2015

5. Benjamins, V.R., Casanovas, P., Breuker, J., Gangemi, A. (eds.): Law and the Semantic Web: Legal Ontologies, Methodologies, Legal Information Retrieval, and Applications. LNCS (LNAI), vol. 3369. Springer, Heidelberg (2005)

6. Boella, G., Di Caro, L., Humphreys, L., Robaldo, L., Rossi, R., van der Torre, L.: Eunomos, a legal document and knowledge management system for the web to provide relevant, reliable and up-to-date information on the law. Artificial Intelligence and Law to appear (2016)

7. Boella, G., Di Caro, L., Graziadei, M., Cupi, L., Salaroglio, C.E., Humphreys, L., Konstantinov, H., Marko, K., Robaldo, L., Ruffini, C., Simov, K., Violato, A., Stroetmann, V.: Linking legal open data: breaking the accessibility and language barrier in European legislation and case law. In: Proceedings of the 15th International Conference on Artificial Intelligence and Law. ICAIL 2015, pp. 171–175. ACM, New York (2015)

8. Boella, G., Di Caro, L., Rispoli, D., Robaldo, L.: A system for classifying multi-label text into eurovoc. In: Proceedings of the Fourteenth International Conference on Artificial Intelligence and Law. ICAIL 2013, pp. 239–240. ACM, New York (2013)

9. Copestake, A., Flickinger, D., Pollard, C., Sag, I.A.: Minimal recursion semantics: an introduction. Res. Lang. Comput. 3(2), 281–332 (2005)

10. Davidson, D.: The logical form of action sentences. In: Rescher, N. (ed.) The Logic of Decision and Action. University of Pittsburgh Press, Pittsburgh (1967)

11. De Hert, P., Papakonstantinou, V., Kamara, I.: The cloud computing standard ISO/IEC 27018 through the lens of the EU legislation on data protection. Comput. Law Secur. Rev. 32(1), 16–30 (2016)

12. Dimyadi, J., Governatori, G., Amor, R.: Evaluating legaldocml and legalruleml as a standard for sharing normative information in the AEC/FM domain. In: Proceedings of the Lean and Computing in Construction Congress (LC3) (to appear, 2017)

13. Giurgiu, A., Lommel, G.: A new approach to EU data protection. Crit. Q. Legislation Law 97(1), 10–27 (2014)

14. Governatori, G., Olivieri, F., Rotolo, A., Scannapieco, S.: Computing strong and weak permissions in defeasible logic. J. Philos. Logic 42(6), 799–829 (2013). http://dx.doi.org/10.1007/s10992-013-9295-1

15. Governatori, G., Rotolo, A., Sartor, G.: Deontic defeasible reasoning in legal interpretation. In: Atkinson, K. (ed.) The 15th International Conference on Artificial Intelligence & Law, San Diego, USA (2015)

16. Hansen, J.: Prioritized conditional imperatives: problems and a new proposal. Auton. Agent. Multi-Agent Syst. 17(1), 11–35 (2008)

17. Hobbs, J.R.: Toward a useful notion of causality for lexical semantics. J. Semant. 22, 181–209 (2005)

18. Hobbs, J.R.: Deep lexical semantics. In: Gelbukh, A. (ed.) CICLing 2008. LNCS, vol. 4919, pp. 183–193. Springer, Heidelberg (2008). doi:10.1007/978-3-540-78135-6_16

19. Hobbs, J.: The logical notation: ontological promiscuity. In: Chapter 2 of Discourse and Inference (1998). http://www.isi.edu/~hobbs/disinf-tc.html
20. Horty, J.: Agency and Deontic Logic. Oxford University Press, New York (2001)
21. Horty, J.: Reasons as Defaults. Oxford University Press, New York (2012)
22. Jørgensen, J.: Imperatives and logic. Erkenntnis **7**, 288–296 (1937)
23. Kamp, H., Reyle, U.: From Discourse to Logic: An Introduction to Model-Theoretic Semantics, Formal Logic and Discourse Representation Theory. Kluwer Academic Publishers, Dordrecht (1993)
24. Makinson, D., van der Torre, L.W.N.: Input/output logics. J. Philos. Logic **29**(4), 383–408 (2000)
25. Makinson, D., van der Torre, L.: Permission from an input/output perspective. J. Philos. Logic **32**, 391–416 (2003)
26. McCarthy, J.: Circumscription: A form of nonmonotonic reasoning. Artif. Intell. **13**, 27–39 (1980)
27. van der Meyden, R.: The dynamic logic of permission. J. Logic Comput. **6**, 465–479 (1996)
28. Mihalcea, R., Corley, C., Strapparava, C.: Corpus-based and knowledge-based measures of text semantic similarity. In: Proceedings of the 21st National Conference on Artificial Intelligence. AAAI 2006, vol. 1, pp. 775–780. AAAI Press (2006). http://dl.acm.org/citation.cfm?id=1597538.1597662
29. Parent, X.: Moral particularism in the light of deontic logic. Artif. Intell. Law **19**(2–3), 75–98 (2011)
30. Reding, V.: The upcoming data protection reform for the European Union. Int. Data Priv. Law **1**(1), 3–5 (2011)
31. Robaldo, L.: Independent set readings and generalized quantifiers. J. Philos. Logic **39**(1), 23–58 (2010)
32. Robaldo, L.: Interpretation and inference with maximal referential terms. J. Comput. Syst. Sci. **76**(5), 373–388 (2010)
33. Robaldo, L.: Distributivity, collectivity, and cumulativity in terms of (in)dependence and maximality. J. Logic, Lang. Inf. **20**(2), 233–271 (2011)
34. Robaldo, L., Humphreys, L., Sun, L., Cupi, L., Santos, C., Muthuri, R.: Combining input/output logic and reification for representing real-world obligations. In: Postproceedings of the 9th International Workiung on Juris-informatics. Lecture Notes in Artificial Intelligence (2016)
35. Robaldo, L., Miltsakaki, E.: Corpus-driven semantics of concession: where do expectations come from? Dialogue Discourse **5**(1), 1–36 (2014)
36. Robaldo, L., Sun, X.: Reified input/output logic: Combining input/output logic and reification to represent norms coming from existing legislation. J. Logic Comput. (to appear, 2017)
37. Robaldo, L., Caselli, T., Russo, I., Grella, M.: From Italian text to TimeML document via dependency parsing. In: Gelbukh, A. (ed.) CICLing 2011. LNCS, vol. 6609, pp. 177–187. Springer, Heidelberg (2011). doi:10.1007/978-3-642-19437-5_14
38. Schuler, K.K.: Verbnet: a broad-coverage, comprehensive verb lexicon. Ph.D. thesis, Philadelphia, PA, USA, aAI3179808(2005)
39. Sun, X., Robaldo, L.: On the complexity of input/output logic. J. Appl. Logic (to appear, 2017)
40. Vibert, H., Jouvelot, P., Pin, B.: Legivoc - connectings laws in a changing world. J. Open Access Law **1**(1), 165–174 (2013)

Mobile Radio Tomography: Agent-Based Imaging

K. Joost Batenburg[1,2], Leon Helwerda[3], Walter A. Kosters[3(✉)],
and Tim van der Meij[3]

[1] Mathematical Institute, Leiden, The Netherlands
[2] CWI, Amsterdam, The Netherlands
[3] Leiden Institute of Advanced Computer Science (LIACS),
Leiden, The Netherlands
w.a.kosters@liacs.leidenuniv.nl

Abstract. Mobile radio tomography applies moving agents that perform wireless signal strength measurements in order to reconstruct an image of objects inside an area of interest. We propose a toolchain to facilitate automated agent planning, data collection, and dynamic tomographic reconstruction. Preliminary experiments show that the approach is feasible and results in smooth images that clearly depict objects at the expected locations when using missions that sufficiently cover the area of interest.

Keywords: Radio tomography · Intelligent agents · Wireless signal strength measurements · Image reconstruction · Localization and mapping

1 Introduction

Radio tomography is a technique for measuring the signal strength of low-frequency radio waves exchanged between *sensors* around an area, and reconstructing information about objects in that area. We send a signal between a source and target sensor of a bidirectional *link*. The signal passes through objects that attenuate it, resulting in a detectably weaker signal at the receiving end. This phenomenon makes it possible to determine where objects are located. The typical setup for radio tomography is illustrated in Fig. 1(left), in which the sensors are situated on the boundaries in an evenly distributed manner. Gray lines represent unobstructed links and red lines indicate links attenuated by the object.

Radio tomography has several benefits over other detection techniques. We can see through walls, smoke or other obstacles. The technique does not require objects to carry sensor devices. The radio waves are non-intrusive, with no permanent effects on people. The technique is less privacy-invasive than optical cameras as the possible level of detail is inherently limited due to the nature of the radio waves. It is not possible to accurately identify a person, but we do aim for reconstructed objects that we recognize as such.

© Springer International Publishing AG 2017
T. Bosse and B. Bredeweg (Eds.): BNAIC 2016, CCIS 765, pp. 63–77, 2017.
DOI: 10.1007/978-3-319-67468-1_5

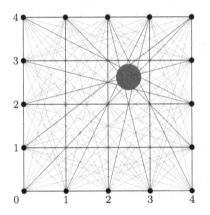

Fig. 1. The sensor network and the physical realization of a vehicle.

A static sensor network with a large number of affordable sensors, placed around an area of interest, can be used to reconstruct and visualize a smooth image in real time [15]. The drawbacks of a static network are the requirement of a large number of sensors and the inability to resolve gaps in the sensor coverage or to react to information obtained through the reconstruction.

One way to resolve these issues is to move the sensors around using *agents*, which are realized as autonomous *vehicles* as pictured in Fig. 1(right). We position them along a *grid* which defines discrete and precise sensor positions. These positions produce a matrix of coordinate-based *pixels* in the reconstructed image. In comparison to the static setup, we require fewer sensors and less prior knowledge about the area. We may adapt the coverage dynamically, for example by zooming in on a part of the area. We name this concept *mobile radio tomography*, which includes both the agent-based measurement collection and the dynamic reconstruction approach.

In this paper, we present our toolchain for mobile radio tomography using intelligent agents, as an engineering effort that builds upon and combines several techniques. In Sect. 2 we describe the key challenges for mobile radio tomography and the components in our toolchain that address them. We then cover two such challenges in greater detail: (i) planning the paths of the agents in Sect. 3, and (ii) reconstructing an image from the measurements in Sect. 4. Results for real-world experiments are presented in Sect. 5, followed by conclusions and further research in Sect. 6. This paper is based on two master's theses on the subject of mobile radio tomography [10,14].

2   Toolchain

Compared to existing localization and mapping techniques that use statically positioned sensors [11,15], radio tomographic imaging with dynamically positioned agents leads to several new challenges. In particular, routes must be planned for each agent such that they obtain isotropic sampling of the network

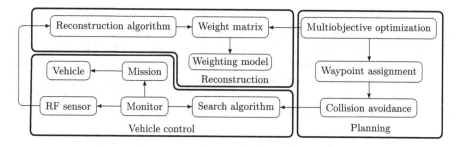

Fig. 2. Diagram of components in the toolchain.

while also shortening the total scanning time and ensuring collision-free movement. Images must be reconstructed from the measurements in real time, requiring algorithms and models that work with a restricted set of data that is potentially incomplete and certainly noisy. Synchronization between the agents must be interleaved with data acquisition using a robust communication protocol.

To deal with these challenges, we developed an open-source, component-based toolchain. The toolchain is mainly written in Python, with low-level hardware components written in C. The diagram in Fig. 2 shows the toolchain's components.

The planning components generate missions for signal strength measurements, as discussed in Sect. 3. The problem of devising a set of link positions to measure is solved by an evolutionary multiobjective algorithm that simulates reconstruction models to ensure that the links cover the entire network. Next, a waypoint assignment algorithm distributes the sensor positions for each link over the vehicles. A path graph search algorithm prevents the vehicles from clashing.

Execution of the mission is taken care of by the vehicle control components. The monitor oversees the process and tracks auxiliary sensors on the vehicles, such as distance sensors for obstacle detection. It makes the RF (radio frequency) sensor perform the signal strength measurements and it may use the search algorithm for collision avoidance during a mission. The mission consists of the waypoints for the vehicle and provides instructions to the vehicle controller. During a mission, this causes the vehicle to move toward the next waypoint.

The reconstruction component converts the signal strength measurements to a two-dimensional visualization of the area of interest. We describe this process in detail in Sect. 4. The weight matrix determines which pixels are intersected by a link, and a weighting model describes how the contents of pixels contribute to a measured signal strength.

3 Missions

We instruct the autonomous vehicles to travel to specific locations around the area of interest, that two-by-two correspond to the positions where sensor measurements must be performed. A vehicle executes its *mission*, which consists of *waypoints* denoting locations to be visited in order. We wish to plan the mission algorithmically instead of assigning waypoints by hand. The vehicles perform measurements together while traversing short and safe paths that do not conflict with concurrent routes. This problem is related to various multi-agent vehicle routing problems with synchronization constraints [6,9,13]. We propose a two-stage algorithm, and describe both parts in text as well as pseudocode.

We assume that we know which links we measure for collecting tomographic data; later on we generate these links using an evolutionary algorithm. To measure a link, sensors must be placed at two positions at the same time. We distribute these tasks over the vehicles. Our assignment algorithm is given as input a set

$$P = \{(p_{1,1}, p_{1,2}), (p_{2,1}, p_{2,2}), \ldots, (p_{\omega,1}, p_{\omega,2})\} \tag{1}$$

with ω location pairs of *coordinate tuples* (two-dimensional vectors), and a set $V = \{v_1, v_2, \ldots, v_\eta\}$ of $\eta \geq 2$ vehicles, initially located at coordinate tuples $S_1, S_2, \ldots, S_\eta$. Now define $U = \{(u, v) \mid u \in V, v \in V, u \neq v\}$, the pairwise unique permutations of the vehicles, e.g., with two vehicles, this is $U = \{(v_1, v_2), (v_2, v_1)\}$.

Our greedy assignment in Algorithm 1 then works as follows: for each vehicle pair $\vartheta = (v_a, v_b) \in U$ and each sensor pair $\rho = (p_{c,1}, p_{c,2}) \in P$, determine the distances $d_1(\vartheta, \rho) = \|S_a - p_{c,1}\|_1$ and $d_2(\vartheta, \rho) = \|S_b - p_{c,2}\|_1$. We use the L^1 norm $\|\cdot\|_1$ to only move in cardinal directions on a grid; in other applications we may use the L^2 norm $\|\cdot\|_2$. Next, take the maximal distance (since one agent must wait for the other to perform a measurement), and finally select the overall minimal pair combination, i.e., solve the following optimization problem:

$$\underset{(\vartheta,\rho) \in U \times P}{\arg \min} \left(\max(d_1(\vartheta, \rho), d_2(\vartheta, \rho)) \right) \tag{2}$$

Algorithm 1. Greedy waypoint assignment

1: **procedure** ASSIGN($S_1, S_2, \ldots, S_\eta, P, V$)
2:     let A_i be a sequence of waypoints for each vehicle v_i, with $i = 1, 2, \ldots, \eta$
3:     $U \leftarrow \{(u, v) \mid u \in V, v \in V, u \neq v\}$
4:     **while** $P \neq \varnothing$ **do**
5:         $\delta_m \leftarrow \infty$
6:         **for all** $(\vartheta, \rho) \in U \times P$ **do** $\triangleright \vartheta = (v_a, v_b)$ and $\rho = (p_{c,1}, p_{c,2})$
7:             $d \leftarrow \max(\|S_a - p_{c,1}\|_1, \|S_b - p_{c,2}\|_1)$
8:             **if** $d < \delta_m$ **then**
9:                 $\delta_m \leftarrow d$, $\vartheta_m \leftarrow \vartheta$ and $\rho_m \leftarrow \rho$
10:            **end if**
11:        **end for** $\triangleright \vartheta_m = (v_a, v_b)$ and $\rho_m = (p_{c,1}, p_{c,2})$
12:        append $p_{c,1}$ to the assignment A_a for vehicle v_a
13:        append $p_{c,2}$ to the assignment A_b for vehicle v_b
14:        $S_a \leftarrow p_{c,1}$ and $S_b \leftarrow p_{c,2}$
15:        remove ρ_m from the set P
16:    **end while**
17:    **return** the assignments $A_1, A_2, \ldots, A_\eta$
18: **end procedure**

The selected positions are then assigned to the chosen vehicle pair, and removed from P. Additionally, S_a becomes the first position and S_b becomes the second sensor position. The greedy algorithm then continues with the next step, until P is empty, thus providing a complete assignment for each vehicle.

Secondly, we design a straightforward collision avoidance algorithm that searches for routes between waypoints that do not *conflict* with any concurrent route of another vehicle; see Algorithm 2. The algorithm is kept simple in order to incorporate it into an evolutionary algorithm (see [7,12] for more intricate methods which result in optimized routes). We use a path graph search algorithm to find a *safe route* that crosses no other routes. Once a vehicle performs a measurement involving another vehicle, their prior routes no longer conflict.

Algorithm 2. Collision avoidance

1: **procedure** AVOID($V, S_1, S_2, \ldots, S_\eta, v_p, v_q, N_p$)
2:     let $W_1, W_2, \ldots, W_\eta$ be sets, with $W_i = \{v_i\}$ for $i = 1, 2, \ldots, \eta$
3:     let G be a graph of discrete positions and connections in the area
4:     remove incoming edges of nodes in G that enter forbidden areas
5:     remove incoming edges of $S_1, S_2, \ldots, S_\eta$ from G
6:     let $R_1, \ldots, R_\eta$ be empty sequences of routes
7:     **for all** $v_i \in V \setminus W_p$ **do**
8:         remove the edges for nodes in R_i from G
9:     **end for**
10:     $R^* \leftarrow$ SEARCH(G, S_p, N_p)   ▷ find a safe path R^* in G from S_p to N_p
11:     append R^* to R_p
12:     reinsert the edges for S_p into the graph G
13:     remove incoming edges for the node N_p
14:     $S_p \leftarrow N_p$ and $W_p \leftarrow W_p \cup \{v_q\}$
15:     **for all** $v_i \in V$ **do**
16:         **if** $v_i \notin W_p$ **then**
17:             reinsert the edges for nodes traversed by the path R_i into G
18:         **end if**
19:         **if** $v_i \neq v_p$ and $W_i = V$ **then**
20:             clear the sequence R_i
21:             $W_i \leftarrow \{v_i\}$
22:         **end if**
23:     **end for**
24:     **return** R_p
25: **end procedure**

Let v_p be the vehicle that we currently assign the position N_p to, and v_q the vehicle that will visit the other sensor position. We also initialize sets $W_1, W_2, \ldots, W_\eta$, where each W_i indicates with which other vehicles the given vehicle v_i has recently performed a measurement. We assume that the search algorithm is given as input a graph G, start point S_p and end point N_p, and outputs a route of *intermediate* points R^*, or an empty sequence if there is no safe path.

We use the collision avoidance algorithm every time the waypoint assignment algorithm assigns a position to a vehicle, so twice per step. Thus, we detect problematic situations as they occur, which are either solved via detours (although the vehicle might also search for a faster safe path while the mission takes place), or by rejecting the entire assignment. The resulting assignments should be collision-free, assuming that all vehicles follow their assigned route and wait for each other at synchronization points, where they also perform their signal strength measurements.

In order to supply the waypoint assignment algorithms with a non-static set of sensor positions P (see (1)), we utilize an evolutionary multiobjective algorithm [8]. The iterative algorithm generates a set of positions and alters it in such a way that it theoretically converges toward an optimal assignment. We keep a population $(X_1, X_2, \ldots, X_\mu)$ of multiple *individuals*, each of which contains variables that encode the positions in an adequate form. After a random initialization, the algorithm performs iterations in which it selects a random individual X_i and slightly mutates it to form a new individual [2].

In our situation, the variables of an individual encode coordinates for positions around the area of interest, and possibly inside of it as well. Define $m^{(i)}$ as the number of pairs of positions that are correctly placed, such that the link between the positions intersects the network. Using these positions, we can deduce other information, such as a weight matrix $A^{(i)}$ (containing link influence on pixels; see Sect. 4), for each individual X_i. The algorithm then removes an individual that is infeasible according to the domain of the variable or due to the *constraints* in (3) and (4), such as a minimum number of valid links ζ. The constraints are wrapped into a combined feasibility value in (5):

$$Q_1^{(i)} : \exists j : \forall k : A_{j,k}^{(i)} \neq 0 \tag{3}$$

$$Q_2^{(i)} : m^{(i)} \geq \zeta \tag{4}$$

$$f_i = \begin{cases} 0 & \text{if } \neg Q_1^{(i)} \vee \neg Q_2^{(i)} \\ 1 & \text{if } Q_1^{(i)} \wedge Q_2^{(i)} \end{cases} \tag{5}$$

In the case that all constraints and domain restrictions are met by each individual, the multiobjective algorithm uses a different selection procedure based on the *objective functions*. We remove an individual if its objective values are strictly higher than those of one *dominating* individual in the population. If none of the individuals are dominated, we remove the one with the minimum crowding distance [5]. The crowding distance is defined as the area around the individual within the objective space. We can place the objective values in this space as a plotted function, which is known as the *Pareto front*.

We provide two objective functions that the evolutionary multiobjective algorithm should minimize. Certain parts of the algorithm favor two over more than two objectives, which is why we combine related functions as terms of one overarching objective. Maximization problems, such as achieving optimal coverage area with the generated links, are converted to minimization objectives by negating them. The objective functions in (6) and (7) describe desirable properties for intersecting links and minimized distances, respectively:

$$g_1(X_i) = -\sum_{j=1}^{m^{(i)}} \sum_{k=1}^{n} A_{j,k}^{(i)} \tag{6}$$

$$g_2(X_i) = \delta \cdot \left(\sum_{j=1}^{m^{(i)}} \left\| p_{j,1}^{(i)} - p_{j,2}^{(i)} \right\|_2 \right) + (1-\delta) \cdot T^{(i)} \tag{7}$$

In the entire selection step of the evolutionary algorithm, we use the reconstruction, waypoint assignment and collision avoidance algorithms to check that a new individual adheres to the constraints and to calculate the objective values. Aside from the link weight matrix $A^{(i)}$ for one individual X_i, we calculate the pairwise L^2 norms between sensor positions, and $T^{(i)}$, the sum of minimized route distances (2), weighted by a factor δ. These algorithms generate missions that provide sufficient network coverage. When we stop the evolutionary multi-objective algorithm, we can manually select one of the individuals and use the mission it generates, using the Pareto front as a reference for balanced objective values [5].

4   Reconstruction

The reconstruction phase takes care of converting a sequence of signal strength measurements to a two-dimensional image of size $m \times n$ pixels that may be visualized. Let $M = \{(s_1, t_1, r_1), \ldots, (s_k, t_k, r_k)\}$ be the input, in which s_i and t_i are pairs of integers indicating the x and y coordinates on the grid for the source and target sensor i, respectively, r_i is the received signal strength indicator (RSSI) and k is the total number of measurements. We express the conversion problem algebraically as $Ax = b$. Here, b is a column vector of RSSI values $(r_1, r_2, \ldots, r_k)^T$, x is a column vector of $m \cdot n$ pixel values (in row-major order) and A is a weight matrix that describes how the RSSI values are to be distributed over the pixels that are intersected by the link, according to a weighting model.

In general, signal strength measurements contain a large amount of noise due to *multipath interference*. The wireless sensors send signals in all directions, and thus more signals than those traveling in the line-of-sight path may reach the target sensor, causing interference. This phenomenon is especially problematic in indoor environments due to reflection of signals. Although techniques exist to include an estimate of the contribution of noise inside the model [15], we suppress noise outside the model using calibration measurements and regularization algorithms. The difficulty lies in the fact that the reconstruction problem is an ill-posed inverse problem, which we have to solve for highly noisy and unstable measurements.

The *weight matrix* A defines the mapping between the input b and the output x. If and only if the contents of a pixel with index i attenuate a link with index ℓ, then the weight $w_{\ell,i}$ in row ℓ and column i is nonzero. Given a link ℓ, a *weighting model* determines which pixels have an influence on this link and are thus assigned nonzero weights. The weights may be normalized using the link length d_ℓ to favor shorter links [16]. The variable $d_{\ell,i}$ is the sum of distances from the center of pixel i to the two endpoints of link ℓ.

The *line model* in (8) assumes that the signal strength is determined by objects on the line-of-sight path, as shown in Fig. 3a; the *ellipse model* in (9) is based on the definition of Fresnel zones:

$$w_{\ell,i} = \begin{cases} 1 & \text{if link } \ell \text{ intersects pixel } i \\ 0 & \text{otherwise} \end{cases} \tag{8}$$

$$w_{\ell,i} = \begin{cases} 1/\sqrt{d_\ell} & \text{if } d_{\ell,i} < d_\ell + \lambda \\ 0 & \text{otherwise} \end{cases} \tag{9}$$

$$w_{\ell,i} = e^{-(d_{\ell,i}-d_\ell)^2/2\sigma^2} \tag{10}$$

Fresnel zones, used to describe path loss in communication theory, are ellipsoidal regions with focal points at the endpoints of the link and a minor axis diameter λ. Only pixels inside this region are assigned a nonzero weight, as seen in Fig. 3b.

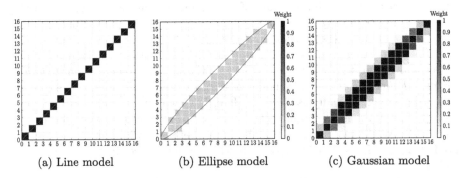

(a) Line model   (b) Ellipse model   (c) Gaussian model

Fig. 3. Illustration of weight assignment for a link from (0, 0) to (16, 16) by the weighting models.

Moreover, we introduce a new *Gaussian model* in (10). This model is based on the assumption that the distribution of noise conforms to a Gaussian distribution. The log-distance path loss model is a signal propagation model that describes this assumption as well [1].

The general Gaussian function is defined as $f(x) = \alpha e^{-(x-\mu)^2/2\sigma^2}$. In this equation α is the height of the curve's peak, μ is the location of the peak's center and σ is the standard deviation that controls the width of the top of the curve. The Gaussian model uses the Gaussian function to assign weights for the pixels. Most weight is assigned to pixels on the line-of-sight path and less weight is assigned to pixels that are farther away, depending on their distance from the line-of-sight path $(d_{\ell,i} - d_\ell)$. We specifically use a Gaussian function with $\alpha = 1$ and $\mu = 0$ since this ensures that pixels on the line-of-sight path get the highest weight, as depicted in Fig. 3c. The parameter σ may be tuned as in practice there appears to be a wide range of suitable values.

Due to the ill-posed nature of the problem, in general there exists no exact solution for $A\boldsymbol{x} = \boldsymbol{b}$ because A is not invertible. Instead, we attempt to find a solution $\boldsymbol{x}_{min}$ that minimizes the error using *least squares approximation* [3] as defined in (11), where $R(x)$ is a regularization term:

$$\boldsymbol{x}_{min} = \arg\min_{\boldsymbol{x}} \left(\|A\boldsymbol{x} - \boldsymbol{b}\|_2^2 + R(\boldsymbol{x}) \right) \qquad (11)$$

The *singular value decomposition* (*SVD*) of A may be used to solve this and is defined as $A = U\Sigma V^T$, in which U and V are orthogonal matrices and Σ is a diagonal matrix with singular values [16]. If we use the exact variant of singular value decomposition which does not apply any regularization, then we have the regularization term $R(\boldsymbol{x}) = 0$. *Truncated singular value decomposition* (*TSVD*) is a regularization method that only keeps the τ largest singular values in the SVD and is defined as $A = U_\tau \Sigma_\tau V_\tau^T$ [16]. Small singular values have low significance for the solution and become erratic when taking the reciprocals for Σ. Moreover, the truncated singular value decomposition is faster to compute, which is an important property for reconstructing images in real time.

While the dimensionality reduction from TSVD does stabilize the solution, the resulting images may still contain unstable spots. Iterative regularization methods incorporate desired characteristics of the reconstructed images. *Total variation minimization* (*TV*, see [16]) enforces that the reconstructed images are smooth, i.e., that the differences between neighboring pixels are as small as possible, by favoring solutions that minimize variability in the resulting image. The gradient $\nabla \boldsymbol{x}$ of $\boldsymbol{x}$ is a measure of the variability of the solution. The regularization term in (11) is set to $R(\boldsymbol{x}) = \alpha \sum_{i=0}^{\xi-1} \sqrt{(\nabla \boldsymbol{x})_i^2 + \beta}$, in which ξ is the number of elements in $\nabla \boldsymbol{x}$. The parameter α indicates the importance of a smooth solution and leads to a trade-off as high value indicates more noise suppression, but less correspondence to the actual measurements. The term is not squared, so we need an optimization algorithm for the minimization. The parameter β is a small value that prevents discontinuity in the derivative when $\boldsymbol{x} = 0$, as that generally needs to be supplied to optimization algorithms.

Finally, let us consider another measure of variability. *Maximum entropy minimization* (*ME*) smoothens the solution by minimizing its entropy. Entropy is a concept in thermodynamics that provides a measure of the amount of disorder in a structure. The *Shannon entropy* is defined as $H = -\sum_{i=0}^{\gamma-1} q_i \log_2(q_i)$, in which γ is the number of unique gray levels in the solution and q_i is the probability that gray level i occurs in the solution. Low entropy indicates a low variation in gray levels (which we observe as noise). While this regularization technique is well-known [4], we have found no previous work discussing its application to radio tomographic imaging. The regularization term in (11) is set to $R(x) = \alpha H$. We calculate a numerical approximation of the derivative. The described reconstruction methods and weighting models allow us to obtain a clear image of the area.

5   Experiments

To study the effectiveness of our approach, we perform a series of experiments. Two vehicles drive around on the boundaries of a 20×20 grid in an otherwise empty experiment room. We use hand-made missions that apply common patterns used in tomography, such as fan beams. With this setup we create a dataset with two persons standing in the middle of the left side and in the bottom right corner of the network, and a dataset with one person standing in the top right corner of the network. A separate dataset is used for calibration.

The first experiment compares all combinations of regularization methods and weighting models to determine which pair yields the most accurate reconstructions. We use the dataset with the two persons, so both must be clearly visible. The outcomes of this experiment are presented in Fig. 4, in which darker pixels indicate low attenuation and brighter pixels indicate high attenuation.

The first observation is that SVD indeed leads to major instabilities because of the lack of regularization. Noise is amplified to an extent that the images do not provide any information about the positions of the persons. TSVD, while being a relatively simple regularization method, provides more stable resulting images that clearly show the positions of the two persons. The ellipse model and the new Gaussian model yield similar clear results, whereas using the line model leads to slightly more noise compared to the former two. Even though TV and ME use different variation measures, the reconstructions are visually the same and equally clear.

Besides providing a clear indication of where the persons are located inside the network, it is important that the reconstructed images are smooth. The second experiment studies the smoothening effects of the regularization methods using 3D surface plots of the raw grayscale images, i.e., without any additional coloring steps applied. The only difference between the experiment runs are the regularization method, so any other parameters remain the same, such as the Gaussian weighting model and the precollected dataset with the two persons that we use as input. The results for this experiment are shown in Fig. 5.

The ideal surface plot consists of a flat surface with two spikes exactly at the positions of the persons. The surface plot for SVD is highly irregular, which leads to noticeable noise in the image due to a high variance in pixel values. In contrast, the surface plot for TSVD is smooth and the two spikes are clearly distinguishable. However, there are still some small unstable spots. The surface plots for TV and ME are, again, practically the same and have even fewer instabilities.

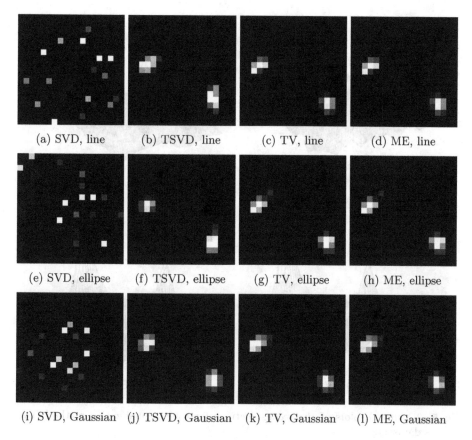

(a) SVD, line   (b) TSVD, line   (c) TV, line   (d) ME, line

(e) SVD, ellipse   (f) TSVD, ellipse   (g) TV, ellipse   (h) ME, ellipse

(i) SVD, Gaussian   (j) TSVD, Gaussian   (k) TV, Gaussian   (l) ME, Gaussian

Fig. 4. Reconstructions combining regularization methods and weighting models (two persons dataset).

With regard to the algorithmically generated missions discussed in Sect. 3, we perform a parameter optimization by comparing the average objective values of the resulting individuals in multiple runs of the evolutionary algorithm. The twelve parameters influence the sensitivity and scale of the optimization algorithm, as well as the waypoint assignment and collision avoidance algorithms. Certain features, such as a mutation operator specially designed to optimize link positions, can be enabled and disabled this way as well. We provide 350 unique combinations of values to these parameters, and repeat each experiment five times.

We find that some of these variables influence the performance in terms of stability, convergence speed and finding optimized positions. For example, the specialized mutation operator finds individuals that have better objective values, but result in chaotic populations over time. The population size hardly affects the algorithm's effectiveness nor speed. Other parameters produce missions which are applicable only if we change the dimensions of the area of interest.

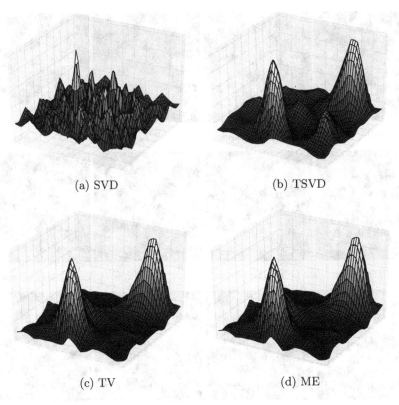

(a) SVD   (b) TSVD

(c) TV   (d) ME

Fig. 5. 3D surface plots of the reconstructions for each regularization method (two persons dataset).

For the final experiment, we generate a mission for a 20×20 grid using the evolutionary multiobjective algorithm, and compare it to a hand-made mission used previously. The algorithm is tuned to place sensors for at least 320 and up to 400 valid links to be measured during the mission. The results of the parameter optimization are applied as well. At 7000 iterations, we end the generation run and pick a *knee point* solution that optimizes both objectives in the resulting Pareto front. A manual check using the collision avoidance algorithm determines that this assignment is safe. In Fig. 6, we show the images resulting from the tomographic reconstruction of the dataset with one person.

The reconstruction, which is run in real time during the collection of signal strength measurements, uses TV and the Gaussian model. We can track the time it takes before a mission provides a smooth and correct result, in terms of quality and realism. In Fig. 6a, we are around halfway through the planned mission, with 204 out of 382 measurements collected. The reconstructed image clearly shows the person standing in the top right corner. Figure 6b shows the reconstructed image provided by the hand-made mission after 413 out of 800 measurements. Although this mission's movements are less erratic (and more measurements

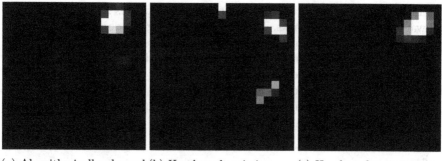

(a) Algorithmically planned (b) Hand-made mission,   (c) Hand-made mission,
mission, halfway result   halfway result   final result

Fig. 6. Reconstructions for algorithmically planned and hand-made missions (one person dataset).

per time unit are made), it is not stable enough to clearly show one person while it develops. Figure 6c shows the end result, where the hand-made mission does provide an acceptable reconstructed image. The planned mission does not diverge from its initial smooth image and we can stop the mission early.

6   Conclusions and Further Research

We propose a mobile radio tomography toolchain that collects wireless signal strength measurements using dynamic agents, which are autonomous vehicles that move around with sensors. We plan missions, which describe the locations that the agents must visit and in what order. Novel algorithms provide us with generated missions, guaranteeing that two sensors are at the right locations to perform a measurement. The algorithms avoid conflicts between the routes and provide an optimized coverage of the network.

The measurements are passed to the reconstruction algorithms to create a visualization of the area of interest that corresponds to the patterns in the data as best as possible. Regularization methods suppress noise in the measurements and increase the smoothness of the resulting image. We introduce a new Gaussian weighting model and apply maximum entropy minimization to the problem of radio tomographic imaging. Preliminary experiments show that the mobile radio tomography approach is effective, i.e., it is able to provide smooth reconstructed images in a relatively short time frame using algorithmically planned missions.

There is much potential for further research. One interesting topic is to replace the agents, that currently operate on the ground using small-scale robotic rover cars, with drones that fly at different altitudes. This leads to 3D reconstruction, e.g., by performing a reconstruction at different altitudes and combining the images, which are slices of the 3D model. The reconstruction algorithms could be modified to allow performing measurements anywhere in the 3D space, although this makes the problem more difficult to solve in real time.

Finally, improvements could be made to the algorithms related to planning and waypoint assignment tasks. This includes altering the objectives of the evolutionary multiobjective algorithm and using predetermined patterns that we encode in the variables. It is also not entirely clear yet how the network coverage, or the lack thereof, influences the quality of the reconstruction. The waypoint algorithm could be rebalanced to take less greedy steps or to factor in the time that certain actions take, such as turning around. Altering missions dynamically helps making adaptive scanning a viable approach.

References

1. Andersen, J.B., Rappaport, T.S., Yoshida, S.: Propagation measurements and models for wireless communications channels. IEEE Commun. Mag. **33**, 42–49 (1995)
2. Bäck, T.: Evolutionary Algorithms in Theory and Practice: Evolution Strategies, Evolutionary Programming, Genetic Algorithms. Oxford University Press, Oxford (1996)
3. Björck, A.: Numerical Methods for Least Squares Problems. SIAM, Philadelphia (1996)
4. Bovik, A.C.: Handbook of Image and Video Processing. Academic Press, New York (2005)
5. Branke, J., Deb, K., Dierolf, H., Osswald, M.: Finding knees in multi-objective optimization. In: Yao, X., et al. (eds.) PPSN 2004. LNCS, vol. 3242, pp. 722–731. Springer, Heidelberg (2004). doi:10.1007/978-3-540-30217-9_73
6. Bredström, D., Rönnqvist, M.: Combined vehicle routing and scheduling with temporal precedence and synchronization constraints. Eur. J. Oper. Res. **191**, 19–31 (2008)
7. De Wilde, B., ter Mors, A.W., Witteveen, C.: Push and rotate: Cooperative multi-agent path planning. In: Proceedings of the 2013 International Conference on Autonomous Agents and Multi-agent Systems, pp. 87–94. International Foundation for Autonomous Agents and Multiagent Systems (2013)
8. Emmerich, M., Beume, N., Naujoks, B.: An EMO algorithm using the hypervolume measure as selection criterion. In: Coello Coello, C.A., Hernández Aguirre, A., Zitzler, E. (eds.) EMO 2005. LNCS, vol. 3410, pp. 62–76. Springer, Heidelberg (2005). doi:10.1007/978-3-540-31880-4_5
9. Golden, B.L., Raghavan, S., Wasil, E.A.: The Vehicle Routing Problem: Latest Advances and New Challenges. Springer, Heidelberg (2008). doi:10.1007/978-0-387-77778-8
10. Helwerda, L.: Mobile radio tomography: Autonomous vehicle planning for dynamic sensor positions. Master's thesis, LIACS, Universiteit Leiden (2016)
11. Menegatti, E., Zanella, A., Zilli, S., Zorzi, F., Pagello, E.: Range-only SLAM with a mobile robot and a wireless sensor networks. In: Proceedings of the 2009 IEEE International Conference on Robotics and Automation, pp. 8–14 (2009)
12. Sharon, G., Stern, R., Felner, A., Sturtevant, N.R.: Conflict-based search for optimal multi-agent pathfinding. Artif. Intell. **219**, 40–66 (2015)
13. Standley, T.S.: Finding optimal solutions to cooperative pathfinding problems. In: Proceedings of the Twenty-Fourth AAAI Conference on Artificial Intelligence, pp. 173–178 (2010)

14. Van der Meij, T.: Mobile radio tomography: reconstructing and visualizing objects in wireless networks with dynamically positioned sensors. Master's thesis, LIACS, Universiteit Leiden (2016)
15. Wilson, J., Patwari, N.: Radio tomographic imaging with wireless networks. IEEE Trans. Mob. Comput. **9**, 621–632 (2010)
16. Wilson, J., Patwari, N., Guevara Vasquez, F.: Regularization methods for radio tomographic imaging. In: Virginia Tech Symposium on Wireless Personal Communications (2009)

Combining Combinatorial Game Theory with an α-β Solver for Clobber: Theory and Experiments

Jos W.H.M. Uiterwijk[(✉)] and Janis Griebel

Department of Data Science and Knowledge Engineering,
Maastricht University, Maastricht, The Netherlands
uiterwijk@maastrichtuniversity.nl, janis.griebel@online.de

Abstract. Combinatorial games are a special category of games sharing the property that the winner is by definition the last player able to move. To solve such games two main methods are being applied. The first is a general α-β search with many possible enhancements. This technique is applicable to every game, mainly limited by the size of the game due to the exponential explosion of the solution tree. The second way is to use techniques from Combinatorial Game Theory (CGT), with very precise CGT values for (subgames of) combinatorial games. This method is only applicable to relatively small (sub)games. In this paper, which is an extended version of [7], we show that the methods can be combined in a fruitful way by incorporating an endgame database filled with CGT values into an α-β solver.

We apply this technique to the game of Clobber, a well-known all-small combinatorial game. Our test suite consists of 20 boards with sizes up to 18 squares. An endgame database was created for all subgames of size 8 and less. The CGT values were calculated using the CGSUITE package. Experiments reveal reductions of at least 75% in number of nodes investigated.

1 Introduction

Clobber is a two-player game invented by Albert, Grossman and Nowakowski in 2001 [1]. It is played on a rectangular board, the size of which can be varied. At the start the squares are alternately filled with black and white stones. The first player (Black, also called Left) controls the black stones, the second player (White, also called Right) controls the white stones. Commonly the start position is a filled 8×8 board as shown in Fig. 1.

The players move in alternating order. A move is done by picking one of the own stones and moving it to an orthogonally adjacent square that is occupied by an opponent stone. The opponent stone gets removed from the board ("clobbered"). The player that makes the last move wins. An example move is shown in Fig. 2, where Black has opted to clobber White's stone in the upper left corner.

© Springer International Publishing AG 2017
T. Bosse and B. Bredeweg (Eds.): BNAIC 2016, CCIS 765, pp. 78–92, 2017.
DOI: 10.1007/978-3-319-67468-1_6

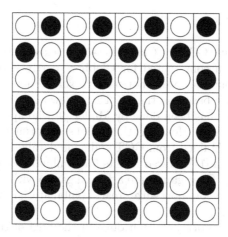

Fig. 1. Common Clobber start position.

Fig. 2. Illustration of a move in Clobber.

Important for Clobber are the following facts:

- Clobber is a 2-player perfect-information, deterministic game.
- Clobber is a converging game. With each move the total number of stones gets decreased by one.
- Clobber is a partizan game. Left and Right have their own stones, which results in the fact, that each player has its own moves.
- If a player can make a move in a specific position, the opponent player could also make a move in that position, because then two opponent stones are orthogonally adjacent. This means the game is a so-called *all-small* game [2].
- The symmetry of a position does not influence the outcome. If the board is mirrored vertically and/or horizontally, or rotated by a multiple of 90°, the outcome of the position does not change.
- No draws are possible: either the first player (Left) or the second player (Right) wins by making the last move.

As a result of these facts Clobber belongs to the category of *combinatorial games*, for which a whole theory has been developed, the Combinatorial Game Theory (CGT). In this article we describe the results of building an endgame database with exact CGT values for the game of Clobber. Section 2 gives an overview of related research. In Sect. 3 we give an introduction to the Combinatorial Game Theory, as far as applicable to Clobber. In Sect. 4 we describe the methods implemented. Next, in Sect. 5, we show our experiments. In Sect. 6 we give our conclusions and some suggestions for further research.

2   Related Work

The literature regarding CGT for Clobber is scarce, except the paper introducing
the game [1]. Claessen constructed for Clobber (partly filled) databases with
exact CGT values, but they were not used to solve boards, but to enhance a
Clobber playing engine based on the MCTS framework [5].

Other related work has been done by Müller [9] who applied CGT values to
solve local endgames in Go. However, the global search is not an α-β search, and
CGT values are not obtained from CGT endgame databases, but calculated on
the spot. Müller and Li [11] showed results for combining α-β search with CGT
pruning and ordering. Their games are artificial games with very special properties, having nothing in common with "real" combinatorial games. No endgame
databases were used.

Most other related work has been done in the area of the game Amazons.
Müller [10] used CGT to establish bounds in a specialized divide-and-conquer
approach. He was able to solve 5×5 Amazons. No CGT endgame databases
were used. Snatzke [13] built CGT endgame databases for a very restricted version of Amazons, namely for subgames fitting on a 2×11 board with exactly
1 queen per player. This was extended in [14] with new results for some small
databases of other shapes, with 1 to 4 queens. He did not incorporate the use of
his databases in a general Amazons solver. Tegos [16] was the first to combine
endgame databases for Amazons in an α-β-based Amazons playing program.
Besides (traditional) minimax endgame databases (without CGT information)
he also implemented CGT endgame databases. These contained just thermograph information, not precise CGT values, and therefore only could be used for
(heuristic) move-ordering purposes. Recently, Song [15] implemented endgame
databases in an Amazons solver. Again, the databases did only contain heuristic (thermograph) information useful for narrowing the bounds in the solving
process. With his program 5×6 Amazons has been solved.

As far as we know the only other research reporting on combining global α-β
searches with endgame databases with precise CGT values is our related work
on Domineering [3,17]. In these articles we showed that equipping a simple α-β
solver with CGT endgame databases gives reductions up to 99% in number of
nodes investigated for solving a testset of 36 non-trivial rectangular boards.

3   Combinatorial Game Theory for Clobber

In this section we give a short introduction to the Combinatorial Game Theory
as far as relevant for Clobber. For a more thorough introduction, we refer to the
literature, in particular [2,4,6].

In a combinatorial game, the players are conventionally called Left and Right.
For Clobber, Left is the player moving the black stones, therefore also denoted
by Black, and similarly Right (White) moves the white stones.

In CGT a game G is represented by its left and right *options* G^L and G^R,
so $G = \{G^L | G^R\}$. In this representation, G^L and G^R stand for sets of games

(the options) that players Left and Right, respectively, can reach by making one move in the game. The *value* of a game indicates how good a game is for a player, where positive values indicate an advantage for Left and negative values an advantage for Right. There are several types of values for Clobber positions. These are treated in the next subsections.

3.1 Numbers

Numbers have the property that any option is a number itself, and that no left option has a higher value than any right option. The simplest number game is the endgame $\{|\}$, denoted as 0. In this position, no player has any available moves, so it is a loss for the player to move. In Clobber, a position with just one stone, either black or white, is an endgame position, with value 0.

Larger or smaller numbers are built recursively:

$$0 = \{|\}$$
$$1 = \{0|\} = \{\{|\}|\}$$
$$2 = \{1|\} = \{\{0|\}|\} = \{\{\{|\}|\}|\}$$
$$-1 = \{|0\} = \{|\{|\}\}$$
$$-2 = \{|-1\} = \{|\{|0\}\} = \{|\{|\{|\}\}\}$$

Also fractions are possible. However, Clobber has the property that no number games are possible, except the endgame 0.

3.2 Nimbers

Nimbers are a class of games where the first player to move wins. The simplest such game is called Star or $*$, defined as $* = \{0|0\} = \{\{|\} | \{|\}\}$, where the player to move has just one option, leading to the endgame. An example is the simple Clobber game of Fig. 3.

Fig. 3. A $*$ position in Clobber.

A *nimber* is a game defined as follows:

$$*0 = 0$$
$$*1 = *$$
$$*2 = \{\{*0, *1\}|\{*0, *1\}\}$$
$$*3 = \{\{*0, *1, *2\}|\{*0, *1, *2\}\}$$
$$*n = \{\{*0, *1, ..., *(n-1)\}|\{*0, *1, ..., *(n-1)\}\}$$

Note that $*0 = 0$ is the only game being both a number and a nimber. All other nimbers have the property that they are a win for the first player to move.

3.3 Ups and Downs

Ups and *downs* are other types of values. An up or $\uparrow$ is defined as $\uparrow = \{0|*\}$ and is strictly positive, meaning that Left wins this game, irrespective of who starts. Down or $\downarrow$ is its negative, defined as $\downarrow = -\uparrow = \{*|0\}$ and is strictly negative, meaning that Right wins this game, irrespective of who starts. See Fig. 4 for an example $\uparrow$ (left) and $\downarrow$ (right) position in Clobber.

Fig. 4. An up and a down position in Clobber.

Next, in Clobber often positions occur that are in fact sums of ups and downs. For these, special notations are introduced, where $\uparrow*$ means $\uparrow + *$, etc.:

$$\uparrow = \{0|*\} \qquad\qquad \downarrow = \{*|0\}$$
$$\uparrow + \uparrow = \Uparrow = \{0|\uparrow*\} \qquad\qquad \downarrow + \downarrow = \Downarrow = \{\downarrow*|0\}$$
$$\uparrow + \uparrow + \uparrow = \text{⇑↑} = \{0|\Uparrow*\} \qquad\qquad \downarrow + \downarrow + \downarrow = \text{⇓↓} = \{\Downarrow*|0\}$$

In general (where $n \cdot \uparrow*$ is parsed as $(n \cdot \uparrow)*$, and similarly for $n \cdot \downarrow*$):

$$n \cdot \uparrow = \{0|*\} \qquad \text{if } n = 1 \qquad n \cdot \downarrow = \{*|0\} \qquad \text{if } n = 1$$
$$n \cdot \uparrow = \{0|(n-1) \cdot \uparrow*\} \quad \text{if } n > 1 \qquad n \cdot \downarrow = \{(n-1) \cdot \downarrow*|0\} \quad \text{if } n > 1$$

They are often combined with a *, giving

$$\uparrow* = \{0, *|0\} \qquad\qquad \downarrow* = \{0|0, *\}$$
$$\Uparrow* = \{0|\uparrow\} \qquad\qquad \Downarrow* = \{\downarrow|0\}$$
$$\text{⇑↑}* = \{0|\Uparrow\} \qquad\qquad \text{⇓↓}* = \{\Downarrow|0\}$$

In general:

$$n \cdot \uparrow* = \{0, *|0\} \qquad \text{if } n = 1 \qquad n \cdot \downarrow* = \{0|0, *\} \qquad \text{if } n = 1$$
$$n \cdot \uparrow* = \{0|(n-1) \cdot \uparrow\} \quad \text{if } n > 1 \qquad n \cdot \downarrow* = \{(n-1) \cdot \downarrow|0\} \quad \text{if } n > 1$$

Figure 5 shows a $\Uparrow*$ (left) and a ⇑↑ (right) position in Clobber.

Fig. 5. Examples of multiple up positions in Clobber.

3.4 Infinitesimal and All-Small Games

Definition 1. *A game G is* infinitesimal *if $-x < G < x$ for every positive number x.*

The number 0 and all nimbers, ups and downs are infinitesimal.

Definition 2. *A game G is* all-small *if for every position H in G it holds that Left can move in H if and only if Right can.*

This means that either $G = 0$ or G^L and G^R are both non-empty and all options are all-small. As a consequence, an all-small game is infinitesimal.

Since Clobber has the property that in every position Black can move if and only if White can, it follows that Clobber is an all-small game. As a consequence all Clobber positions are infinitesimal. So, the only number that can occur in Clobber is 0 (the only number that is also a nimber). Furthermore, only nimbers, ups and downs, and combinations thereof can occur in Clobber.

3.5 Canonical Forms

A game in CGT always has a unique *canonical form*. The canonical form describes the *smallest* game G' of a game G where G' is equivalent to G.

The reduction to the canonical form can be obtained considering two conditions. G' is said to be in canonical form if G' has no *dominated options* and when *reversible options* are *bypassed*. In order to reduce a game to its canonical form, the dominated or reversible options should be identified and removed or bypassed, respectively.

An option is dominated when there is at least one other option equal to or better than the option considered. Then the player has a better or at least an equal alternative. More precisely, a left option L_1 is dominated if there exists another left option L_2 with $L_1 \leq L_2$. A right option R_1 is dominated if there exists another right option R_2 with $R_1 \geq R_2$. A game with a dominated option is equivalent to the smaller game G' with this option removed.

An example Clobber position with dominated options is given in Fig. 6.

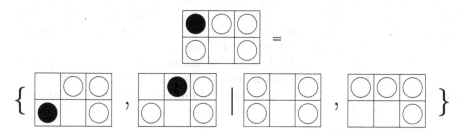

Fig. 6. A Clobber position with dominated options.

The left options have values 0 and $\downarrow$, respectively, the right options both have value 0. Therefore, this position has value $\{0, \downarrow \mid 0, 0\}$. Since 0 dominates $\downarrow$ for Left and 0 dominates 0 for Right, the canonical form of this game is $\{0 \mid 0\} = *$.

An option is reversible if the resulting game contains an option for the next player which is, from its point of view, equal to or better than the current game. In this case it is obvious that the next player would perform this equal or better option immediately after the reversible option. We can then in fact directly use the options of this resulting position instead. We call this *bypassing a reversible option*. More precisely, a left option L_1 of a game G is reversible if L_1 contains a right option $L_1^{R_1}$ with $L_1^{R_1} \leq G$. L_1 can then be replaced by the left options of $L_1^{R_1}$. Vice versa, a right option R_1 of a game G is reversible if R_1 contains a left option $R_1^{L_1}$ with $R_1^{L_1} \geq G$. R_1 can then be replaced by the right options of $R_1^{L_1}$. A game with a reversible option is equivalent to the smaller game G' with this option bypassed.

An example Clobber position with a reversible option is given in Fig. 7.

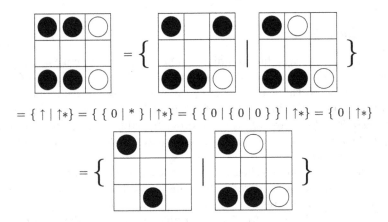

$$= \{ \uparrow \mid \uparrow * \} = \{ \{ 0 \mid * \} \mid \uparrow * \} = \{ \{ 0 \mid \{ 0 \mid 0 \} \} \mid \uparrow * \} = \{ 0 \mid \uparrow * \}$$

Fig. 7. A Clobber position with a reversible option.

In this position we first have removed the dominated options for both players (moves in the lower row being equivalent with moves in the upper row). Next, the left option $\uparrow$ has as only right option a $*$ position and since $*$ definitely is better for Right than the initial position $(\uparrow + \uparrow)$, we may replace the left option in the initial position by the left option of $*$, being 0.

The canonical form helps to show equivalence of games. Since the canonical form of a game is unique, two games are equivalent if their canonical forms are identical. This paper will deal with canonical forms in the endgame database, explained in Sect. 4.

3.6   Sums of Games

Many combinatorial games have the nice property that they can split into sub-games that do not interact. Clobber belongs to this category. When this happens we say that the value of the position is the sum of the values of the subgames. Formally: $G + H = \{G^L + H, G + H^L \mid G^R + H, G + H^R\}$, which states that a player can choose the subgame in which to play.

When the subgames have simple values (0, nimbers, ups and downs) we then can easily add the resulting values. Here a 0 acts as an identity $(0 + G = G)$. Nimbers are added using Nim-addition $(*m + *n = *(m \oplus n))$, where $\oplus$ is the exclusive-or operator applied to the binary representations of the operands. Ups and downs are summed using the rules $m \cdot \uparrow + n \cdot \uparrow = (m + n) \cdot \uparrow$ and $\downarrow = -1 \cdot \uparrow$.

As an example, consider the Clobber position in Fig. 8.

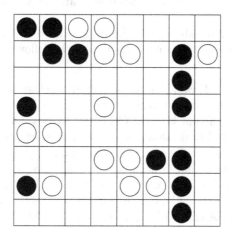

Fig. 8. An example Clobber position with 6 subgames.

The subgames in this example position, from top to bottom and left to right, have values $*2$, $\Uparrow*$, $\downarrow$, 0, $*$, and $*2$. Therefore, the value of the whole position is $*2 + \Uparrow* + \downarrow + 0 + * + *2 = \uparrow$. It means that the position is won by Black, regardless of who is starting the game.

4   Methods

In this section we describe the methods implemented in our Clobber solver. The core is a straightforward α-β solver, described in Sect. 4.1. The main enhancement consists of the construction of the CGT endgame database (Sect. 4.2). Section 4.3 shows how the CGT values from the endgame database can be used within the α-β framework.

4.1   Basic α-β Solver

For the basic Clobber solver we implemented a straightforward zero-window α-β [8] searcher. This searcher investigates lines until the end (returning that either Left or Right made the last move and so wins the game). Since our goal was to measure the impact of using CGT on solution tree sizes, we refrained from incorporating any further enhancements to the α-β framework.

4.2   A CGT Database for Clobber

In order to use CGT values for Clobber subgames we have built a Clobber endgame database with all subgames consisting of 1–8 connected stones with arbitrary shape. Positions with mirror symmetry (left/right and/or top/bottom) are unified into a single entry.

The database is used to store pre-calculated CGT values of specific positions. These CGT values are used to get knowledge about its subgames. If a position is reached that only consists of pre-calculated subgames, the whole position can be solved by combining the CGT values using game-logic rules (when the CGT values are all simple).

As a preparation a database must be designed, created and filled with the appropriate values. These steps are described in the following subsections.

Database design

In the database the CGT values must be stored together with the appropriate game positions. The easiest way to store the CGT values is using a character string because of the different kinds of CGT value types. For the game positions a hash code is generated that allows to rebuild the exact positions. The format of the hash code is described in the next subsection.

Since the information that is needed is simple, only one table is needed for the database. The hash code is used as the primary key. Since the hash code consists of two values, a string and an integer for the position width, the primary key is a combined one. A third column is used for the CGT value of the position.

Representing a position

A game position is represented in a hash code to get stored in the database. It is important that the hash code is unique for a position. There are different approaches that can be used to generate such a unique hash code.

In our first approach the board was iterated sequentially and indicated in the hash code. The squares are visited from left to right through all rows from top to bottom. An empty square is indicated by the character "0", a square with a black stone by "1" and a square with a white stone by "2". These characters are concatenated to a string and used as the hash code for the position. An example is shown in Fig. 9 (left).

Fig. 9. Two example positions with the same hash code and different values.

This position with value $\{*, \uparrow |*, \downarrow\}$ has hash code "211020". A problem with a hash code in this format is that it does not include the position size. The position in Fig. 9 (right) is clearly different, with value $\{0|\downarrow\} = *$, but has the same hash code. To solve this issue the position size needs to be stored together with the hash code. For that the width of a position is added as a column to the database. The primary key consists of the combination of the generated hash code and the position width.

Calculating CGT values

Since the database will be used as a lookup for solved positions, it must be filled with CGT values. The CGT values for the Clobber subgames are calculated with the CGSUITE software tool [12]. These values are in canonical form, which makes them a unique representation of the CGT values. They are stored in the third column in the database.

Generating positions

To fill the database a set of positions is needed for which the appropriate CGT values are calculated and stored. To generate the hash codes for these positions an algorithm is implemented that iterates through all possibilities.

The algorithm gets a limit that consists of a number of connected nonempty squares. All possible formations and fillings of the connected squares are iterated to fill the database. To get all positions with n connected nonempty squares, an empty $n \times n$ board is created. In a naive approach, the algorithm would go recursively through all squares, filling each square first with empty, then with a black stone and at last with a white stone. For an $n \times n$ board it would mean that the algorithm generates $3^{(n \times n)}$ combinations of fillings. These include many combinations that exceed the set limit of connected nonempty squares n.

To avoid iterating through combinations that do not fit the set limit n the algorithm counts the number of nonempty squares through the iterations. If the limit n is reached it continues just with empty squares till the last square of the board. The resulting amount of combinations can be described with binomial coefficients. For an $n \times n$ board it means that there are

$$\binom{n \times n}{n} \times 3^n \tag{1}$$

combinations generated.

As a second improvement the algorithm is changed to ignore combinations that contain less than n nonempty squares. Since the algorithm is executed sequentially with an increasing limit n, it does not need to check such combinations. This improvement is obtained by not generating positions that have less than n nonempty squares reaching the last square of the board. This improvement leads to a smaller amount of evaluated combinations that can be described by the following formula, since always exactly n squares are nonempty:

$$\binom{n \times n}{n} \times 2^n \tag{2}$$

Note that we only consider positions that do not consist of subgames, since it would mean that the amount of nonempty squares is less than n for each subgame separately and therefore will have been generated in earlier iterations.

As a last improvement the amount of iterated positions is drastically decreased. Consider that an $n \times n$ board is used to get all formations of n nonempty connected squares. The width n of the position is only needed if all nonempty squares are in a row. Therefore a height of 1 is sufficient to obtain all possible formations, because there is only one line with all nonempty squares in a row. If one square gets occupied that is not in a row with the others, a smaller position with size $(n - 1) \times 2$ can be used to generate all missing positions. For a next square that is not in row 1 or 2 it means that a position with size $(n - 2) \times 3$ fits to get all possible positions with a maximum of $n - 2$ squares in a row, etc. As an example of generating all possible formations for $n = 5$ connected squares, Fig. 10 shows the required board sizes, where the gray stones just indicate example fillings with five stones, irrespective of their actual colors.

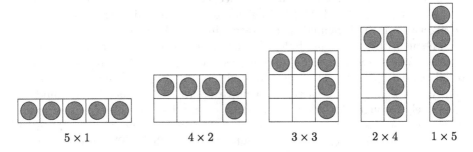

Fig. 10. Board sizes needed to get all possible formations for $n = 5$ connected stones.

We replace the amount of squares, that are not in a row with any other square, but still connected with the other ones, by a variable p. Using boards with size $(n - p + 1) \times p$, where p runs from 1 to n, we can generate all possible formations of n connected nonempty squares. The following formula describes the amount of iterated positions for one specific p:

$$\binom{(n - p + 1) \times p}{n} \times 2^n$$

When we let p run from 1 to n and calculate the sum of the amount of all formation possibilities, the number of iterated positions is described by the following formula:

$$\left(\sum_{p=1}^{n} \binom{(n - p + 1) \times p}{n} \right) \times 2^n \qquad (3)$$

To show the improvement by these approaches for the generation of the positions, the amount of iterated positions is listed in Table 1.

The final improvement results in a serious redundancy reduction. Generating one position means that the hash code is generated, a lookup in the database is

Table 1. Improving position generation.

n	No optimization $3^{(n \times n)}$	n chosen squares Eq. (1)	n nonempty squares Eq. (2)	Final Eq. (3)
1	3	3	2	2
2	81	54	24	8
3	19,683	2,268	672	48
4	4.30×10^7	1.47×10^5	29,120	512
5	8.47×10^{11}	1.29×10^7	1.70×10^6	7,680
6	1.50×10^{17}	1.42×10^9	1.24×10^8	1.45×10^5
7	2.39×10^{23}	1.88×10^{11}	1.10×10^{10}	3.31×10^6
8	3.43×10^{30}	2.90×10^{13}	1.13×10^{12}	8.84×10^7

made, and sometimes a new entry is stored in the database. Now the algorithm only generates 8.84×10^7 instead of 3.43×10^{30} positions for $n = 8$ nonempty connected squares. This is an improvement by a factor 4.05×10^{22}.

Filling the database

The C# algorithm, that generates the positions was prepared to run with one specific set n. To fill the database it first runs with $n = 1$. After completing the generation of all positions, n is increased by 1 and the algorithm is started again. The idea is to get positions generated for an n that is as high as possible. The higher the n is the longer the algorithm runs because the amount of iterated positions increases exponentially.

Running on a 3 GHz Quad-Core computer the algorithm needed around 2 h for all positions with $n \leq 7$ nonempty connected squares. For $n = 8$ it took more than 3 days. Expecting a few months for $n = 9$ the generation was stopped. The number of generated entries for n connected nonempty squares is listed in Table 2. Note that for $n \geq 4$ these numbers are smaller than the ones in the right column of Table 1, since the positions generated may consist of unconnected subfragments already in the database for smaller values of n. The database consumes 79.23 MB memory space.

Table 2. Stored positions in the database.

n	1	2	3	4	5	6	7	8
	2	8	48	304	2,016	13,824	97,280	697,600

sum = 811,082 entries

4.3 Combining the CGT Endgame Database with α-β

Combining the CGT endgame database with an α-β solver can be achieved by a simple adaptation of the basic α-β algorithm. As soon as a (sub)game with at most 8 connected stones is encountered, its value is looked up in the database and using this value it is determined whether the (sub)game is a win or loss for the player to move.

As a result, when investigating an $m \times n$ board lines are never searched deeper than $(m \times n) - 8$, when the position is necessarily reduced to 8 stones (since each move consists of removing exactly 1 stone). Moreover, many lines will even terminate at shallower depths when a position splits into multiple subgames of size ≤ 8. As long as there are subgames with more than 8 stones, the investigation continues. As soon as each subgame separately contains at most 8 stones, the search terminates and game logic is used to determine the winner. This ensures determining correctly the game-theoretical result.

5   Experiments and Discussion

In this section the implemented features are tested to generate insight into their impact on improving a plain α-β search. In the setup we used the α-β implementation, without and with database support, as described in the previous section.

Boards with different sizes are investigated. The size runs up to 18 squares with a maximum width of 8. The board is initially filled in the common way as a checkerboard, starting with a black stone. Since the results are not comparable on different computers considering the investigated time, the number of visited nodes is used as the value to compare.

Two remarks are in order. First, although the efficiency of α-β depends on move ordering, we have not incorporated this in the current implementation. It is, however, far from trivial to find a good move-ordering technique for an all-small game like Clobber. Second, who starts influences the amount of visited nodes. Therefore each position is played twice, once with Black and once with White as the player to move. Both amounts of visited nodes are summed and given as the results shown in Table 3.

In this table all investigated board sizes (with the board between brackets) are listed in the left column. The second column shows the number of visited nodes (disregarding the root in the tree) using the default α-β search (denoted α-β_noDB). The third column contains the number of visited nodes when the database is used (α-β_withDB). Column 4 shows the reductions for α-β_withDB compared to α-β_noDB as percentages. Finally, the last column contains the outcome classes for the boards. Here, $\mathcal{P}$ means a win for the previous player, so a loss for the player to move, $\mathcal{N}$ means a win for the next player (the player to move), whereas $\mathcal{L}$ and an $\mathcal{R}$ denote wins for Left (Black) and Right (White), irrespective who starts the game. The results in this column match with the results in [1] in so far as given there.

One should keep in mind that the filling of each board starts with a black stone and then goes further alternating between White and Black. Therefore, in the position of size 1×7, if the filling would start with a white stone, then Black would be determined as the winner, irrespective of who starts (i.e., outcome class $\mathcal{L}$). All other results just depend on the player to move first and do not change.

Table 3. Comparison of numbers of nodes visited during searches.

Board size	α-β_noDB	α-β_withDB	red. (%)	Outcome class
1 (1×1)	0	0	100.00%	$\mathcal{P}$
2 (1×2)	2	0	100.00%	$\mathcal{N}$
3 (1×3)	2	0	100.00%	$\mathcal{N}$
4 (1×4)	3	0	100.00%	$\mathcal{N}$
4 (2×2)	6	0	100.00%	$\mathcal{N}$
5 (1×5)	15	0	100.00%	$\mathcal{P}$
6 (1×6)	42	0	100.00%	$\mathcal{P}$
6 (2×3)	68	0	100.00%	$\mathcal{P}$
7 (1×7)	49	0	100.00%	$\mathcal{R}$
8 (1×8)	60	0	100.00%	$\mathcal{N}$
8 (2×4)	162	0	100.00%	$\mathcal{N}$
9 (3×3)	222	2	99.10%	$\mathcal{N}$
10 (2×5)	654	66	89.91%	$\mathcal{N}$
12 (2×6)	16,150	1,003	93.79%	$\mathcal{P}$
12 (3×4)	19,532	1,412	92.77%	$\mathcal{P}$
14 (2×7)	45,502	2,849	93.74%	$\mathcal{N}$
15 (3×5)	125,638	14,573	88.40%	$\mathcal{N}$
16 (2×8)	304,230	52,384	82.78%	$\mathcal{N}$
16 (4×4)	625,544	105,596	83.12%	$\mathcal{N}$
18 (3×6)	24,626,986	6,222,980	74.73%	$\mathcal{P}$

6   Conclusions and Future Research

We have shown how CGT values of subgames can be used as an enhancement of a basic α-β solver for Clobber. A database with exact CGT values was built for all (sub)games up to 8 connected stones. Reductions depend on board size, going down from 100% for the boards in the database to 75% for the 3×6 board.

As future research we want to extend the database to include larger subgames. Moreover, we see opportunities to incorporate more game-dependent knowledge, like move-ordering heuristics, to solve larger and more complex Clobber boards. Also, well-known game-independent techniques like transposition tables can greatly enhance the solving efficiency. Of course, it has to be investigated for such optimized α-β solvers for Clobber whether incorporating a CGT endgame database gives similar efficiency enhancements. Finally, we envision using CGT theory in solvers for other combinatorial games (besides Domineering [3] and Clobber (this work)) to get more insight into the game conditions for such a fruitful combination.

References

1. Albert, M.H., Grossman, J.P., Nowakowski, R.J., Wolfe, D.: An introduction to Clobber. Integers Electron. J. Comb. Number Theor. **5**(2), 12 p. (2005)
2. Albert, M.H., Nowakowski, R.J., Wolfe, D.: Lessons in Play: An Introduction to Combinatorial Game Theory. A K Peters, Wellesley (2007)
3. Barton, M., Uiterwijk, J.W.H.M.: Combining combinatorial game theory with an α-β solver for Domineering. In: Grootjen, F., Otworowska, M., Kwisthout, J. (eds.), BNAIC 2014: Proceedings of the 26th Benelux Conference on Artificial Intelligence, pp. 9–16. Radboud University, Nijmegen (2014)
4. Berlekamp, E.R., Conway, J.H., Guy, R.K.: Winning Ways for your Mathematical Plays, vol. 1–2. Academic Press, London (1982). 2nd edition, in four volumes: vol. 1 (2001), vols. 2, 3 (2003), vol. 4 (2004), A K Peters. Wellesley
5. Claessen, J.: Combinatorial game theory in Clobber. Master's thesis, Maastricht University (2011)
6. Conway, J.H.: On numbers and games. Academic Press, London (1976). 2nd edition A K Peters, Natick, MA (2001)
7. Griebel, J., Uiterwijk, J.W.H.M.: Combining combinatorial game theory with an α-β solver for Clobber. In: Bosse, T., Bredeweg, B. (eds.), BNAIC 2016: Proceedings of the 28th Benelux Conference on Artificial Intelligence, pp. 48–55. University of Amsterdam/Vrije Universiteit Amsterdam, Amsterdam (2016)
8. Knuth, D.E., Moore, R.W.: An analysis of alpha-beta pruning. Artif. Intell. **6**, 293–326 (1975)
9. Müller, M.: Global and local game tree search. Inf. Sci. **135**, 187–206 (2001)
10. Müller, M.: Solving 5 × 5 Amazons. In: The 6th Game Programming Workshop (GPW 2001), Hakone (Japan), 2001, vol. 14, IPSJ Symposium Series, pp. 64–71 (2001)
11. Müller, M., Li, Z.: Locally informed global search for sums of combinatorial games. In: van den Herik, H.J., Björnsson, Y., Netanyahu, N.S. (eds.) CG 2004. LNCS, vol. 3846, pp. 273–284. Springer, Heidelberg (2006). doi:10.1007/11674399_19
12. Siegel, A.N.: Combinatorial game suite: a computer algebra system for research in combinatorial game theory (2003). http://cgsuite.sourceforge.net/
13. Snatzke, R.G.: Exhaustive search in the game Amazons. In: Nowakowski, R.J. (ed.) More Games of No Chance, Proceedings of MSRI Workshop on Combinatorial Games, Berkeley, CA, July 2000, MSRI Publ., vol. 42, pp. 261–278. Cambridge University Press, Cambridge (2002)
14. Snatzke, R.G.: New results of exhaustive search in the game Amazons. Theoret. Comput. Sci. **313**, 499–509 (2004)
15. Song, J.: An enhanced solver for the game of Amazons. Master's thesis, University of Alberta (2013)
16. Tegos, T.: Shooting the last arrow. Master's thesis, University of Alberta (2002)
17. Uiterwijk, J.W.H.M., Barton, M.: New results for Domineering from combinatorial game theory endgame databases. Theoret. Comput. Sci. **592**, 72–86 (2015)

Aspects of the Cooperative Card Game Hanabi

Mark J.H. van den Bergh[1,2], Anne Hommelberg[1], Walter A. Kosters[1(✉)],
and Flora M. Spieksma[2]

[1] Leiden Institute of Advanced Computer Science,
Universiteit Leiden, Leiden, The Netherlands
w.a.kosters@liacs.leidenuniv.nl
[2] Mathematical Institute, Universiteit Leiden, Leiden, The Netherlands

Abstract. We examine the cooperative card game HANABI. Players can only see the cards of the other players, but not their own. Using hints partial information can be revealed. We show some combinatorial properties, and develop AI (Artificial Intelligence) players that use rule-based and Monte Carlo methods.

1 Introduction

The game of HANABI, meaning "fire flower" or "fireworks" in Japanese, is a cooperative card game that requires the players to combine efforts in order to achieve the highest possible score. Designed by Antoine Bauza in 2011 and published by R & R Games [8] (see Fig. 1), among others, it is designated as a game in the categories "cooperative play" and "hand management" by the game analysis site BOARDGAMEGEEK [3]. The goal of the game is simple: play out several sequences of cards in the right order. The catch, however, is that the players can only see the cards in other player's hands and not their own; information has to be gathered by a system of hints that reveal partial information.

A game in the named categories that is somewhat similar in gameplay, is THE GAME designed by Steffen Benndorf [3]. Though leading to interesting combinatorial and strategic questions, there is not much to be found in the literature on this kind of games. More is known about another game which bears similarity to Hanabi in view of its dealing with hints: BRIDGE. Yet another game which gives rise to theoretical problems comparable to those inspired by Hanabi, because of its similarity in nature, is SOLITAIRE; as an example, NP-completeness results for HANABI can be found in [1].

In [5] the authors describe two sophisticated strategies that play the game in a near perfect way. They make clever use of the hints system, using dedicated conventions, and codings as in the well-known hats problem (cf. [6]). Note that, as in Bridge, conventions may be artificial: the natural (literal) meaning of the hints might be or is (as in [5]) even ignored. In our paper we restrict ourselves to this natural meaning only.

This paper examines some interesting properties of the game, but is far from being complete. In fact, we show complicated mathematical behavior for a one-player version without hints, and provide an exploratory examination of some artificial players (cf. [2,7]), including rule-based and Monte Carlo versions.

© Springer International Publishing AG 2017
T. Bosse and B. Bredeweg (Eds.): BNAIC 2016, CCIS 765, pp. 93–105, 2017.
DOI: 10.1007/978-3-319-67468-1_7

Fig. 1. HANABI, as sold by R & R Games [8].

We start with a comprehensive explanation of the rules of HANABI in Sect. 2. We then deal with two main questions. In Sect. 3, we consider the playability of a game of HANABI: given a certain start configuration, is it possible to obtain a maximal score when playing perfectly? We will address some theoretical issues. Next, in Sect. 4, we look for a good strategy for any arbitrary game. Among others, we consider Monte Carlo methods, which seem promising. In Sect. 5, we conclude with a summary of the given results as well as some interesting open questions.

2   Game Rules

The classic game of HANABI is played with a stack of $N = 50$ cards. Every card has one out of $C = 5$ colors — blue (B), red (R), green (G), yellow (Y) or white (W) — and a value between 1 and $k = 5$. In the classic game, for every color there are three 1s, two 2s, 3s and 4s and one 5, hence fifty cards in total. At the start of the game, the stack is shuffled and a hand of cards is dealt to each player: $R = 5$ cards are given to every player if $P = 2$ or $P = 3$ persons are playing, and $R = 4$ cards are dealt in games with $P = 4$ or $P = 5$ players. Now, every player picks up his/her cards in such a way that the other players can see them, but they themselves cannot. The rest of the cards forms the face-down stack.

The goal of the game is to create C stacks of cards going from 1 through k on the table, one of each color. To do so, players take turns, choosing exactly one of the three following actions every turn:

– give a hint,
– discard a card,
– play a card.

At the start of the game, a pool of $H = 8$ hint tokens is available to the players. To give a hint to another player, one token must be removed from this pool.

If there are no tokens left, this action cannot be chosen. Giving a hint is done by either pointing out all cards (perhaps zero) of a certain color or all cards of a certain value in the hand of one other player. This is explained most easily by considering an example hand like (R3, W1, B1, R4, B5). A hint may be expended to point out the position of the W1 and B1, telling the player that *these two cards are 1s*. One might also point out the R3 and R4 by telling that *these cards are red*. Also, pointing out B5 by saying *this is a 5* is fine. However, one cannot point out B5 by telling that *this card is blue*, because B1 must then also be pointed at. One is allowed to tell the player that *there are no green cards in the hand*, as this hint effectively points out all (zero) green cards. This option is sometimes disallowed, as in [5].

The second possible action is to discard a card: the active player takes a card from his/her hand (without first looking at it) and announces that it will be discarded. It is then put face-up in the discard pile, which may be viewed by all players at any point in the game. Once in the discard pile, a card will never re-enter the game. A new card is now taken from the face-down stack to replenish the hand and as an added bonus, a hint token is added to the available hints pool. Therefore, if there are already H hints available, this action cannot be chosen.

Finally, one may choose to try to play a card much in the same way as discarding a card. However, the players now look whether the card can be added to one of the stacks on the table. This is again best clarified by an example. Suppose the following stacks are on the table:

$$
\begin{array}{|c|c|c|}
\hline
\text{R1} & \text{G1} & \text{W1} \\
\text{R2} & & \text{W2} \\
& & \text{W3} \\
\hline
\end{array}
$$

It is now possible to play a R3, which can be appended to the leftmost stack. Similarly, a G2 or W4 would be fine. It is also possible to play a Y1 or B1, starting a new stack. However, one may for example not play a R4 (as a R3 is first needed) or R2 (as this is already on the table). Moreover, another R1, G1 or W1 cannot be played as there may only be one stack of every color on the table at any time. If a card is successfully played, it is put in the fitting position on the table. If a card turns out not to be playable, it is moved to the discard pile and the players score one error. In either case, the active player draws a new card from the stack, but no hint token is added.

In order to show that there is more to a hint than just its plain meaning, consider the following example. If a player has only one 2, the G2, a hint pointing at this card might in this situation be considered containing the message "come on, play this card".

The game ends when one of three conditions is met. First, if $E = 3$ errors are made by trying to play a card, the game ends with the lowest possible score of 0. Second, if the C-th stack is completed by playing a card, i.e., if there is a sequence of 1 through k of every color on the table, the game ends with the highest possible score of $C \cdot k$ (for the classic game: 25). Finally, when the last card of the face-down stack is drawn, every player gets one more turn, including the player that drew the last card. The achieved score is then determined by adding the highest number of every stack on the table. In the previous example, a score of $2 + 1 + 3 = 6$ points would be obtained. Apart from just aiming at the highest score, one might also focus on the 0–1 target of obtaining the maximal score.

Naturally, the rules for classic HANABI can easily be generalized and extended. One can for example alter the parameters mentioned above, see Sect. 3. A more formal approach can be found in [2].

3   Playability

In this section, we will address the first main question: given an initial configuration of a game of HANABI, is it possible to obtain the maximum score if playing perfectly? An initial configuration of a game of HANABI (or simply a game of HANABI) is called *playable* if the maximum score can be achieved. We first turn to a theoretical approach involving combinatorics; a perhaps more practical approach using dynamic programming is presented in [1] and also in [2].

As in [1], we consider the simplified situation where the players can also see their own cards (making the hints system and errors obsolete). We give a result for the one-player version, with $R = 1$, called SINGLEHANABI: a player must immediately play or discard the newly received card. We consider only one color, but the number of cards of each value can be arbitrary.

Fix an integer $k \geq 1$. Let $x = (x_1, x_2, \ldots, x_k)$ be a vector with k non-negative integers $x_1, x_2, \ldots, x_k$, and let $F(x) = F(x_1, x_2, \ldots, x_k)$ denote the number of ordered sequences of length $N = N(x) = x_1 + x_2 + \ldots + x_k$ with x_1 occurrences of the integer 1, x_2 occurrences of the integer 2, $\ldots$, x_k occurrences of the integer k, *without* a subsequence 1–2–$\ldots$–k: the so-called *bad* sequences; the others are called *good*. Note that a subsequence is not necessarily consecutive; e.g., the sequence 1–3–2–3 has 1–3–2 and 1–2–3 as subsequence, but not 2–3–1. So $F(x_1, x_2, \ldots, x_k)$ is the number of unplayable SINGLEHANABI games, having x_i cards of value i ($1 \leq i \leq k$): there is no subsequence 1–2–$\ldots$–k that would allow the player, who has "no memory", to immediately play these k "cards" in order.

If for some i with $1 \leq i \leq k$ we have $x_i = 0$, then we know that $F(x) = \binom{N}{x_1, x_2, \ldots, x_k, N - \sum_{j=1}^{k} x_j}$, since all sequences are bad in that case.

We also note that F is symmetric in its arguments. Indeed, we construct a bijection between good sequences with interchanged numbers of, e.g., 1 s and 2 s: $x_1 \leftrightarrow x_2$. To do this, view every good sequence in parts: the part up to but not including the first 1, the part between this 1 and the next 2, etc. Now, the required bijection is given by swapping the first and second part and changing all 1 s in the parts to 2 s and vice versa. We clarify this by an example, in which the numbers which divide the parts are shown in bold:

$$2413134131\mathbf{3}21332421421 \leftrightarrow 323423231\mathbf{1}422331412412$$

Note that the resulting sequence contains the proper amount of every number. Moreover, applying the proposed bijection twice results in the original sequence, which shows that it is indeed bijective.

Inspired by a discussion by anonymous contributors at STACKEXCHANGE [10], our main result in this section is:

Theorem. We have

$$F(x) = \sum_{\substack{y \prec x \\ |y| \leq k-2}} a(N, y).$$

Here we denote $y \prec x$ if the ordered sequence y with k non-negative integers satisfies $y_i \leq x_i$ for all i with $1 \leq i \leq k$. Furthermore, $|y|$ denotes the number of non-zero elements in y. We put $a(n, y) = (-1)^{|y|} \binom{n}{y}^* (k - |y| - 1)^{n - s(y)}$. The multinomial coefficient $\binom{n}{y}^*$ is defined as follows: $\binom{n}{y}^* = \binom{n}{y_1 \ominus 1, y_2 \ominus 1, \ldots, y_k \ominus 1, n - s(y)} = n!/((y_1 \ominus 1)!(y_2 \ominus 1)! \ldots (y_k \ominus 1)!(n - s(y))!)$ (the bottom last term in the multinomial coefficient, here $n - s(y)$, is often omitted by convention) with $s(y) = \sum_{i=1}^{k}(y_i \ominus 1)$, where we used $t \ominus 1 = \max(t - 1, 0)$.

The equation can also be written as

$$F(x) = \sum_{\ell=0}^{k-2}(-1)^\ell \sum_{\substack{y \prec x \\ |y| = \ell}} \binom{N}{y}^* (k - \ell - 1)^{N - s(y)}.$$

For example, with $k = 3$ and $\ell = |y| = 1$, in the computation of $F(2, 3, 1)$ we encounter sequences $(1, 0, 0)$, $(2, 0, 0)$, $(0, 1, 0)$, $(0, 2, 0)$, $(0, 3, 0)$ and $(0, 1, 0)$; and, e.g., $\binom{6}{(0,2,0)}^* = \binom{6}{0,1,0,5} = \binom{6}{1,5} = \binom{6}{1} = 6$. The term with $\ell = |y| = 0$ equals $(k - 1)^N$. Note that $F(x_1, x_2)$ evaluates to 1, as expected. By the way, $F(x_1)$ is 0, if $x_1 > 0$.

Proof. The theorem can be proven through the obvious recurrence (for $x_1 > 0$)

$$F(x_1, x_2, \ldots, x_k) = \binom{N - 1}{x_1 - 1} F(x_2, \ldots, x_k) +$$
$$+ F(x_1, x_2 - 1, \ldots, x_k) + \ldots + F(x_1, x_2, \ldots, x_k - 1)$$

where we interpret a term as 0 if one of its arguments is negative. The respective terms count bad sequences that start with a 1, with a 2, ..., with a k.

We first note that $F(1, \overbrace{0, \ldots, 0}^{k-1 \text{ times}})$ equals 0 if $k = 1$, and equals 1 if $k > 1$. Furthermore, $F(0, 0, \ldots, 0) = 1$. This is the basis for an inductive proof, with respect to $\langle k, N \rangle$.

Using the symmetry of F in its arguments, we may assume that $x_1 > 0$. If $x_2 > 0$ we compute (analogous for the other terms):

$$F(x_1, x_2 - 1, \ldots, x_k) = \sum_{\substack{y \prec x; y_2 < x_2 \\ |y| \leq k-2}} a(N - 1, y) = \sum_{\substack{y \prec x; y_2 = 0 \\ |y| \leq k-2}} a(N - 1, y) + \sum_{\substack{y \prec x; y_2 \neq 0 \\ |y| \leq k-2}} a(N, y) \frac{y_2 - 1}{N}$$

Now we look at a fixed y, and combine all contributions of the $k - 1$ terms. If $y_1 = 0$ we arrive at $a(N - 1, y)(k - |y| - 1) + a(N, y)s(y)/N = a(N, y)$. However, if y has non-zero y_1, we have to be more careful. We then still arrive at $a(N, y)$, but now with an additional term $a(N - 1, y) - a(N, y)(y_1 - 1)/N$. If we let y_1 increase from 1 to x_1 (where we keep the other elements from y unchanged), these terms telescope to $a(N - 1, y')$ with $y'_1 = x_1$ and $y'_i = y_i$ $(1 < i \leq k)$. We rewrite

$$\binom{N - 1}{x_1 - 1} F(x_2, \ldots, x_k) = - \sum_{\substack{y \prec x; y_1 = x_1 \\ |y| \leq k-2}} a(N - 1, y),$$

which exactly cancels the remaining terms.

If we happen to have $x_2 = 0$, the argument above remains valid. Indeed,

$$\sum_{\substack{y \prec x \\ |y| \leq k-2}} a(N - 1, y) = \sum_{\substack{y \prec x; y_1 < x_1 \\ |y| \leq k-2}} a(N - 1, y) + \sum_{\substack{y \prec x; y_1 = x_1 \\ |y| \leq k-2}} a(N - 1, y)$$

is equal to 0, since the first term from the right hand side in that case equals

$$F(x_1 - 1, 0, x_3, \ldots, x_k) = \binom{N - 1}{x_1 - 1, x_3, \ldots, x_k},$$

whereas the second equals

$$-\binom{N - 1}{x_1 - 1} F(0, x_3, \ldots, x_k) = -\binom{N - 1}{x_1 - 1}\binom{N - x_1}{x_3, \ldots, x_k} = -\binom{N - 1}{x_1 - 1, x_3, \ldots, x_k}.$$

thereby completing the proof.   □

One consequence is that

$$F(\overbrace{1, 1, \ldots, 1}^{k \text{ times}}) = k! - 1 = \sum_{\ell=0}^{k-2} (-1)^\ell \binom{k}{\ell} (k - \ell - 1)^k$$

which happens to be a special case of a formula from [9]. In this same category, another special case is

$$F(\overbrace{4, 4, \ldots, 4}^{13 \text{ times}}) = \sum_{\ell=0}^{11} (-1)^\ell \binom{13}{\ell} \sum_{i_1=0}^{3} \cdots \sum_{i_\ell=0}^{3} \binom{52}{i_1, \ldots, i_\ell} (12 - \ell)^{52-i_1-\ldots-i_\ell}$$

which evaluates to 9197327002632448456557941835019448965558291 2237019 (a prime number approximately equal to $9 \cdot 10^{49}$) using a straightforward Python program, that takes a few seconds. Therefore, the probability that a shuffled deck of standard playing cards contains a full increasing subsequence (from ace to king, disregarding suits) turns out to be equal to $1 - F(4, 4, \ldots, 4)/(52!/4!^{13}) \approx 0.000554$, which is a folklore result.

4   Strategies

We will now consider the second main question: what is a good strategy? In this section, a *strategy* is a means to determine which action to take in any given state of the game. We will look at games with $P = 3$ players in which every player has a hand size of $R = 5$ as in the classic game rules in Sect. 2. We call a card *useful* if it can be appended to one of the stacks on the table at the current moment. A card is called *worthless* if it is clear that it can never played (either because a copy of it has already been successfully played (called "dead" in [5]) or all copies of a lower numbered card of the same suit have been discarded).

In [7], strategies are considered in which the players try to estimate their hand by analyzing actions of other players. This analysis is for the two-player version, and relies heavily on online learning. The author claims an average result of 15.85 points. Here, we will try two other types of strategies, and explore their potential. The first one implements several rules of thumb that tend to come up quickly in human play: a rule-based strategy. The second is an implementation of a basic Monte Carlo strategy. Note that it is hard to determine the maximal score that can be achieved for any given game, see also [1]. However, the results from [5] suggest that a perfect score is realizable in most cases.

In contrast with [5], we will not consider conventions regarding the hints. Effectively, hints will be taken literally. In a game state, we will only use hint information on individual cards, and, for instance, not use the identity of the player who provided a particular hint. Clearly, this makes the hints system less powerful.

For the first strategy, every player acts according to the following preset rules:

1. If there is a card in my hand of which I am "certain enough" that it can be played, I play it.
2. Otherwise, if there is a card in my hand of which I am "certain enough" that it is worthless, I discard it.
3. Otherwise, if there is a hint token available, I give a hint.
4. Otherwise, I discard a card.

In this framework, there are several parameters to be determined. First, we may choose the definition of "certain enough" in steps 1 and 2: we let $\omega_p, \omega_d \in [0, 1]$ be the thresholds above which a player knowing that the probability of a card being useful, resp. worthless, exceeds the threshold, it is played, resp. discarded, in step 1, resp. 2. Moreover, one can choose whether or not to take any risk to end the game on three errors by prohibiting to play cards which are not certainly useful after having made two errors; this is referred to as "safe play".

In step 3, we can follow different guidelines which determine the hint to be given. We consider four of them: (#1) random; (#2) giving a hint that gives information on the largest number of cards; (#3) giving a hint on the next useful card in sight or on the largest number of cards if no useful card is seen; or (#4) giving a hint on the next useful card or on the next worthless card if no useful card is available or otherwise on the largest number of cards.

Finally, in step 4, we choose from four different rules by which the card to be discarded is chosen. There are: (#1) random; (#2) discarding the card of which it is most certain that it is worthless; (#3) discarding the card which has been stored in hand the longest; or (#4) discarding the card of which it is most certain that it is not absolutely necessary to complete all stacks (a card which is not "indispensable", according to [5]).

In addition to these choices, we also explore whether it is profitable to sometimes swap the order of steps 3 and 4. We let $\omega_h \in [0, 1]$ be the probability that this is done during a turn.

To test the various settings of the parameters, we let three players using the same strategy play 10,000 different starting configurations. Every configuration is played ten times in order to account for the randomness in the strategies. A test run of this kind takes approximately one minute on a computer with an Intel i7 2.3 GHz core and 6 GB of RAM.

First, we take $\omega_d = \omega_h = 1.0$, and vary ω_p in $\{0.5, 0.6, 0.7\}$. The average scores for all sixteen combinations of hint and discard rules are shown in Table 1. It is apparent that the third hint rule, giving a hint on the next useful card or otherwise on the largest number of cards, is dominant for this setting of the parameters. Surprisingly, discard rules #1 and #2 are about as effective, meaning that discarding randomly is competitive with discarding the card of which we most think it is worthless.

Now, we vary the other parameters. For a full overview of the results, see [2]. It turns out that for the combination of hint rule #3 and discard rule #2, the best results can be obtained. In Fig. 2 the results for varying ω-thresholds can be found, playing safely: no more risk is taken when playing cards after two errors

Table 1. Scores obtained with different rules and varying ω_p, for $\omega_d = \omega_h = 1.0$, playing safely.

		ω_p	Hint rule			
			#1	#2	#3	#4
Discard rule	#1	0.7	6.3	13.5	14.5	11.9
		0.6	7.1	13.9	15.3	12.7
		0.5	7.3	13.8	15.2	12.5
	#2	0.7	6.6	13.8	14.5	12.0
		0.6	7.4	14.4	**15.4**	12.7
		0.5	7.5	14.2	15.3	12.6
	#3	0.7	5.8	13.1	14.1	11.5
		0.6	6.6	13.5	14.9	12.3
		0.5	6.9	13.4	14.8	12.3
	#4	0.7	5.8	13.0	13.9	11.5
		0.6	6.7	13.7	14.8	12.3
		0.5	6.8	13.5	14.7	12.2

have been made. We see (also in Table 1) that the best average score obtained is 15.4 when taking $\omega_p = 0.6$, $\omega_d = \omega_h = 1.0$ and playing safely. Apparently, it is profitable to try and play a card once we are 60% sure that it will be correct (unless we have already made two errors in which case we require certainty), and only discard a card if it is certain that it is worthless and give a hint if possible otherwise.

For the Monte Carlo strategy, the basis is well-known. In every turn, we try every action after which the game is played out by random players many times. Each of these random games is evaluated in some way, after which we choose to do the action which led to the best score.

In the implementation for HANABI, special care has to be taken in at least two situations. First, note that by trying to play a card and seeing how this turns out, a player could illegitimately obtain information on this card. Therefore, when trying to play a card in the Monte Carlo phase, the hand of the active player is shuffled through the deck and a new hand is dealt which is consistent with all hint information obtained so far. This way, the factual information on the cards is stored without allowing the agent to cheat. However, information on the exact hints that were given and the time at which these were given is lost, which possibly results in the agent not picking up implied hints, e.g., a hint pointing out a card actually meaning that it can successfully be played. This is in sharp contrast with the inner working of the strategies as described in [7]. Still, much progress could be booked for this determinization step by better judging or even learning the probability distribution of the cards in the hand of the active player, incorporating the hints provided so far.

Second, a truly random player will end the game on three errors with high probability. To circumvent this, we choose to not end the game after three errors have been made in the play-out phase. Moreover, we prohibit the random player from playing a card of which it is certain that it cannot be played based on the hints received so far. If the random player knows that none of the cards in his/her hand can be played, he/she will always randomly discard a card or give a hint. If he/she knows that only some of his/her cards are useless, he/she may also randomly try to play one of the other cards.

Contrary to the standard implementation of Monte Carlo, we do not evaluate each of the play-outs on the final score obtained. Instead, we register for the next D turns the amount of new points obtained as well as the amount of errors made with respect to the current turn in each play-out. For each new point we administer a $+1$ and for the k-th error we administer a score of $-k$. If at least three errors were made, we administer an extra penalty of -2 (this value was chosen based on experimentation). Experiments showed that without this additional penalty, the average score of the Monte Carlo player improves slightly, but the number of times the game ends with a score of 0 due to three errors being made, increases as well. We then compute a weighted sum over the scores for each of the D turns, counting later turns with (linear) higher weight, after which we pick the action with the highest average score among the play-outs. Note that taking the value $D = 1$ would result in greedy play.

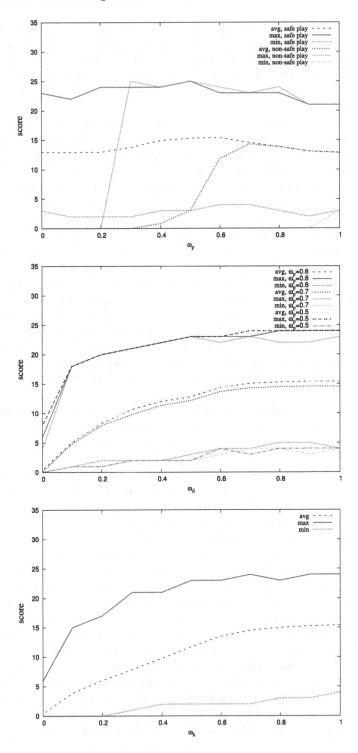

Fig. 2. Scores obtained using hint rule #3 and discard rule #2, for varying ω-thresholds.

Furthermore, it turns out to be profitable to not let the random player choose an action uniformly at random. Indeed, a Monte Carlo player using this strategy achieves on average a score of approximately 8.5 (with representative values like 1000 play-outs and $D = 5$); note that such a player is unable to differentiate between possible hint moves.

The enhanced random player first decides whether to play, discard or give a hint, with probability 1/6, 1/6 and 2/3, respectively (the chosen values may benefit from further tuning). If one or more of these action types cannot be chosen at this time, the probabilities of selecting the remaining action types keep the same proportion to each other. For example, if the player knows none of his/her cards can be played, the probability of choosing a discard and hint action will be 1/5 and 4/5 respectively. If the random player decides to play a card, he/she favors playing a card which he/she deems likely to be playable, based on the number of possible cards that still remain in play and still conform to the information gained through hints. Similarly, when discarding a card, the random player favors a card which is likely to remain useless for the rest of the game, and favors keeping those cards that are likely to be the last remaining copy of that card and need to still be played later in the game to achieve the maximum score. Finally, hints are chosen uniformly at random.

In order to test the performance of the Monte Carlo player, we let three players using this strategy play 500 different starting configurations. A test run of this kind takes approximately three hours on a computer with an Intel i7 2.3 GHz core and 6 GB of RAM, when using 1000 play-outs and a depth of 10.

In the experiments, the results of which can be seen in Fig. 3, we have varied the amount of turns D that the Monte Carlo player takes into account and the number of play-outs allowed per different action. Surprisingly, the value of D taken, when not being too small, does not fundamentally affect the obtained score. Furthermore, while raising the amount of allowed play-outs does seems to somewhat improve the average score obtained, the differences are relatively small.

In fact, we note that it is not unlikely that these differences are to be contributed for a great part to random chance. Indeed, the standard deviation in the scores obtained seems to be quite high, with scores as low as 6 (or sometimes even 0) and as high as 24 being observed with an average around 15 points; the highest average score obtained is 16, with 1000 play-outs. We also see this large range in the scores obtained by the rule-based strategy discussed earlier in this section. Apparently, as the results from [5] show, substantial improvements can be reached by using more than the literal meaning of the hints. As discussed in the previous section and in [1], it is sometimes impossible to score a perfect game with 25 points. An easy example is a game where many 1 s and/or 2 s arrive late in the deck.

A provisional conclusion could be that, with the current settings, rule-based players and Monte Carlo players show similar performance, where the latter perform a little better. The results from [7] for the two-player version, using online learning, even seem to be a little better than for the three-player version.

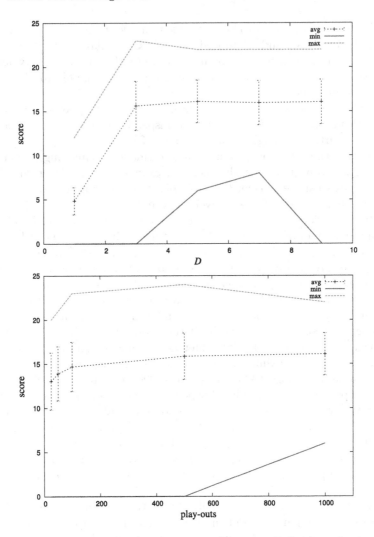

Fig. 3. Scores obtained using 1000 play-outs and varying D (top), and using $D = 5$ and varying number of play-outs (bottom), both with standard deviation.

5   Conclusions and Further Research

In this paper we have examined some interesting aspects of the cooperative card game HANABI, that has the special property that players can only see the cards in the hands of the other players — but not in their own. Hints provide partial information about the game states; in contrast with [5] we only consider the literal meaning of hints.

Even simplified versions of the game, like SINGLEHANABI, give rise to complicated combinatorial questions and corresponding formulas. Furthermore, we have shown that different game playing strategies offer promising results. In particular, Monte Carlo techniques require little game knowledge, but are capable of delivering high quality competitive players. One issue is the problem of the amount of information that is used during the play-outs. The simple rule-based players perform a little worse, even incorporating more game knowledge.

As further research we first mention the quest for a proof of the result from Sect. 3 using the principle of inclusion and exclusion, perhaps also providing links to more general theorems. It is also of interest to find (substantial) classes of games that are unplayable. For the Monte Carlo players, we like to combine the presented method with techniques as mentioned in [4], like UCT and/or information sets, meanwhile somehow learning knowledge. And finally, there is much potential in research regarding the hints system, as the results from [5] show; it clearly pays off to examine clever hint schemes.

Acknowledgements. The authors would like to thank Hendrik Jan Hoogeboom, as well as the anonymous contributors from [10]. Moreover, they thank the reviewers for bringing [5] to our attention.

References

1. Baffier, J.-F., Chiu, M.-K., Diez, Y., Korman, M., Mitsou, V., van Renssen, A., Roeloffzen, M., Uno, Y.: Hanabi is NP-complete, even for cheaters who look at their cards. In: Proceedings of 8th International Conference on Fun with Algorithms (FUN 2016), Leibniz International Proceedings in Informatics (LIPIcs) 49, pp. 4:1–4:17 (2016)
2. van den Bergh, M.J.H.: Hanabi, a cooperative game of fireworks. Bachelor thesis, Leiden University (2015). www.math.leidenuniv.nl/scripties/BSC-vandenBergh. pdf
3. BoardGameGeek, website. www.boardgamegeek.com. Accessed 3 Oct 2017
4. Browne, C., Powley, E., Whitehouse, D., Lucas, S., Cowling, P.I., Rohlfshagen, P., Tavener, S., Perez, D., Samothrakis, S., Colton, S.: A survey of Monte Carlo Tree Search methods. IEEE Trans. Comput. Intell. AI Games 4, 1–43 (2012)
5. Cox, C., De Silva, J., Deorsey, P., Kenter, F.H.J., Retter, T., Tobin, J.: How to make the perfect fireworks display: two strategies for Hanabi. Math. Mag. 88, 323–336 (2015)
6. van Ditmarsch, H., Kooi, B.: One Hundred Prisoners and a Light Bulb. Springer, Switzerland (2015)
7. Osawa, H., Hanabi, S.: Estimating hands by opponent's actions in cooperative game with incomplete information. In: Proceedings of the Workshop at the Twenty-Ninth AAAI Conference on Artificial Intelligence: Computer Poker and Imperfect Information, pp. 37–43 (2015)
8. R & R Games, website. www.rnrgames.com. Accessed 3 Oct 2017
9. Ruiz, S.: An algebraic identity leading to Wilson's theorem. Math. Gaz. 80, 579–582 (1996)
10. StackExchange, website Sequences that contain subsequence 1,2,3. http://www.math.stackexchange.com/questions/1215764/sequences-that-contain-subsequence-1-2-3. Accessed 3 Oct 2017

Solving the Travelling Umpire Problem
with Answer Set Programming

Joost Vennekens[(⊠)]

Department of Computer Science @ Technology Campus De Nayer, KU Leuven,
J.-P. De Nayerlaan 5, 2860 St-katelijne-waver, Belgium
joost.vennekens@cs.kuleuven.be

Abstract. In this paper, we develop an Answer Set Programming (ASP) solution to the Travelling Umpire Problem (TUP). We investigate a number of different ways to improve the computational performance of this solution and compare it to the current state-of-the-art. Our results demonstrate that the ASP solution is superior to other declarative solutions, such as Constraint Programming, but that it cannot match the most recent special-purpose algorithms from the literature. However, when compared to the earlier generation of special-purpose algorithms, it does quite well.

1 Introduction

The Travelling Tournament Problem is a well-known optimisation problem, where the goal is to schedule a series of games at different venues such that the travel distance for the participating teams is minimized. The *Travelling Umpire Problem (TUP)* is a related problem, in which the games and their location are given and the goal is to assign an umpire to each of these games. This assignment must satisfy a number of hard constraints (e.g., the same umpire should not referee the same team two weeks in a row) and the overall distance travelled by the umpires should be minimized.

This problem, which abstracts a real-life task for the Major League Baseball (MLB), was first introduced into the scientific literature in 2007 [15]. Over the last five years, it has received a good deal of attention within the scientific community [4,5,13,14,16–18,20,21]. One of its interesting characteristics is that the TUP has a compact description, but is computationally very challenging. Its associated decision problem is NP-complete [5].

In this paper, we investigate whether it is possible to solve the TUP using the declarative paradigm of Answer Set Programming (ASP). For this, we will make use of the state-of-the-art ASP solver Clasp [8], which is a regular winner of the ASP competition [1–3,6,9,10]. As we will show below, the TUP can be represented in ASP in an elegant and modular representation. Such a representation has the advantage that it can easily be extended with additional constraints, in order to better capture the real-life MLB problem. The computational performance of our method is superior to that of other declarative paradigms, and

T. Bosse and B. Bredeweg (Eds.): BNAIC 2016, CCIS 765, pp. 106–118, 2017.
DOI: 10.1007/978-3-319-67468-1_8

matches that of the first special-purpose algorithm to be developed for TUP [16]. However, it cannot match the performance of more recent special-purpose algorithms, such as the state-of-the-art method from [14]. Nevertheless, it is of course a benefit of our declarative approach that improvements to the general-purpose Clasp server automatically speed up our approach as well.

This paper is structured as follows. Section 2 recalls the basic concepts of ASP and Sect. 3 the definition of the TUP. In Sect. 4, our basic encoding of TUP in ASP is then presented. Next, Sect. 5 discusses some ways of improving the computational performance of this encoding. We evaluate the performance of our method and compare it to existing approaches in Sect. 6. Finally, Sect. 7 presents our conclusions.

2   Preliminaries: Answer Set Programming

In this section, we briefly recall the Answer Set Programming (ASP) language and it semantics [11].

An *atom* is an expression $p(T_1, \ldots, T_n)$ where p is an n-ary predicate symbol ($n \geq 0$) and the T_i are either constants (starting with a lower case letter) or variables (starting with an upper case letter). A *(normal) rule* is an expression of the form:

$$A \;:- \; B_1, \ldots, B_n, \mathbf{not}\; C_1, \ldots, \mathbf{not}\; C_m. \tag{1}$$

Here, A and all of the B_i and C_j are atoms. The atom A is called the *head* of the rule and $B_1, \ldots, B_n$, $\mathbf{not}\; C_1, \ldots, \mathbf{not}\; C_m$ its *body*. A *(normal) logic program* is a finite set of such rules. An atom, rule or program is called *ground* if it does not contain any variables.

A *belief set* X is a set of ground atoms. Such a set X *satisfies* a ground rule r of form (1) if A belongs to X or there exists an $i \in [1, n]$ such that $B_i \notin X$ or a $j \in [1, m]$ such that $C_j \in X$. A belief set is a model of a ground program P if it satisfies all rules $r \in P$.

For a ground rule r of the form (3) and a belief set X, the reduct r^X is defined whenever there is no atom C_j for $j \in [1, m]$ such that $C_j \in X$. If the reduct r^X is defined, then it is the rule: $A \;:- \; B_1, \ldots, B_n$. The reduct P^X of the ground program P consists of all r^X for $r \in P$ for which the reduct is defined. A belief set X is a *stable model* or *answer set* of P, denoted $X \models_{st} P$, if it is the least model of P^X.

The answer sets of a non-ground program are defined as the answer sets of the *grounding* of the program. This grounding is constructed by replacing each non-ground rule r by the set of all ground rules that can be constructed by replacing all the variables in r by constants in all possible ways.

An answer set solver is a program that computes the stable models of a given program. Typically, these solvers extend the basic ASP language with some additional constructs to make it easier to represent interesting problems. In this paper, we will use the input language of the solver Clasp [7]. This language allows the following additional constructs.

A rule of the form

$$:- B_1, \ldots, B_n, \text{not } C_1, \ldots, \text{not } C_m, \tag{2}$$

is called a *constraint*. It is seen as an abbreviation for a rule

$$f :- B_1, \ldots, B_n, \text{not } C_1, \ldots, \text{not } C_m, \text{not } f. \tag{3}$$

where f is a fresh atom. The effect of such a rule is that no answer set may contain both all of the positive literals B_i and none of the negative literals C_i. A constraint therefore expresses that its body may *not* be satisfied.

Another common pattern in ASP programming is the use of a loop over negation to represent a choice between two alternatives. For instance, to express that either p or q must hold, the following two rules may be used:

```
p :- not q.    q :- not p.
```

The answer sets of this small program are precisely {p} and {q}. However, because this style of programming quickly grows cumbersome, Claps also allows to represent choices as:

```
1 { p, q} 1.
```

This expresses that precisely one of the two atoms p and q must holds (the leftmost 1 states that at least one of these atoms must hold and the rightmost 1 states that at most one may hold). It has the same two answer sets as the above two rules (but the two programs are not strongly equivalent—for that, an additional constraint :- p,q should be added to the two rules).

This construct is made even more powerful by the fact that more complex set expressions may be used instead of a simple enumeration of atoms. For instance, 1{p(X) : q(X)}1 states that precisely one atom $p(X)$ must hold out of all atoms $p(X)$ for which X is such that $q(X)$ holds. If we add to this rule the following three facts:

```
q(1).    q(2).    q(3).
```

the answer sets will be precisely {p(1)}, {p(2)} and {p(3)}. The above three facts may also be conveniently abbreviated as: q(1..3).

3   The Travelling Umpire Problem

The TUP is the problem of assigning n umpires to a double round-robin tournament of $2n$ teams. Since each team plays every other team twice, there are $r = 2(2n - 1)$ different rounds, in which each team plays against exactly one other team. Each game is played at the home venue of one of the two participating teams. The assignment of umpires is subject to three hard constraints:

- Each umpire should visit each team's home venue at least once.
- An umpire should not visit the same home venue more than once in any sequence of q_1 rounds.
- An umpire should not referee the same team more than once in any sequence of q_2 rounds.

The parameters q_1 and q_2 control the tightness of the constraints: higher values are more difficult. Clearly, it only makes sense to have $q_2 \leq q_1 \leq r/2$.

In addition to these three hard constraints, there is also an objective function that must be minimized, namely, the total distance travelled by all of the umpires (assuming they start in the location where their first game is played). To calculate this objective function, a (symmetric) distance matrix is given which enumerates the distances between the home venues of each pair of teams.

Figure 1 shows an example instance for $n = 2$ and a solution for the problem with $q_1 = 2$ and $q_2 = 1$ (note that this problem has no solutions for $q_2 > 1$; taking $q_2 = 1$ means that the third constraint can simply be ignored). The solution shown is also the optimal solution (regardless of which distance matrix is used), since the only other solution that exists (modulo the symmetry between the two umpires) is that in which they needlessly move around between rounds 3 and 4.

Round	Home	Away	Umpire
0	Team 1	Team 2	?
	Team 3	Team 4	?
1	Team 1	Team 3	?
	Team 4	Team 2	?
2	Team 2	Team 3	?
	Team 4	Team 1	?
3	Team 2	Team 1	?
	Team 4	Team 3	?
4	Team 3	Team 1	?
	Team 2	Team 4	?
5	Team 3	Team 2	?
	Team 1	Team 4	?

$\rightarrow$

Round	Home	Away	Umpire
0	Team 1	Team 2	Umpire 1
	Team 3	Team 4	Umpire 2
1	Team 1	Team 3	Umpire 2
	Team 4	Team 2	Umpire 1
2	Team 2	Team 3	Umpire 1
	Team 4	Team 1	Umpire 2
3	Team 2	Team 1	Umpire 2
	Team 4	Team 3	Umpire 1
4	Team 3	Team 1	Umpire 2
	Team 2	Team 4	Umpire 1
5	Team 3	Team 2	Umpire 1
	Team 1	Team 4	Umpire 2

Fig. 1. An instance of the TUP for $n = 2, q_1 = 2, q_2 = 1$ and its optimal solution.

4   Representing the TUP in ASP

Problem instance. The parameters of an instance are represented by a set of ASP facts

limit_big(q_1).   limit_small(q_2).

team($1..2n$).   umpire($1..n$).   round($0..r$).

The tournament schedule is given by a set of facts of the form plays(i, j, r), which denote that home-team i plays away-team j in round r. The distance matrix is represented by a set of facts distance(i, j, δ), indicating that the distance between the home venue of team i and that of team j is δ. Only facts for $i \leq j$ are included. The symmetric information is represented by a rule

```
dist(X,Y,D) :- dist(Y,X,D).
```

Finally, it is also convenient to include a unary predicate last($r - 1$) which indicates the number $r - 1$ of the last round and a unary predicate max_dist(δ_{max}) which indicates the largest number that occurs in the distance matrix.

The example shown in Fig. 1 corresponds to the following set of facts:

```
limit_big(2).    plays(2,3,2).   dist(1,2,745).  dist(3,3,0).
limit_small(1).  plays(4,1,2).   dist(1,3,665).  dist(3,4,380).
team(1..4).      plays(2,1,3).   dist(1,4,929).  dist(4,1,929).
umpire(1..2).    plays(4,3,3).   dist(2,1,745).  dist(4,2,337).
round(0..5).     plays(3,1,4).   dist(2,2,0).    dist(4,3,380).
plays(1,2,0).    plays(2,4,4).   dist(2,3,80).   dist(4,4,0).
plays(3,4,0).    plays(3,2,5).   dist(2,4,337).  max_dist(929).
plays(1,3,1).    plays(1,4,5).   dist(3,1,665).  last(5).
plays(4,2,1).    dist(1,1,0).    dist(3,2,80).
```

Generating the search space. We represent the possible solutions to the problem by a set of atoms move(u, t, r), meaning that umpire u moves to the home venue of team t in round r. For instance, move(1,1,0) means that umpire 1 moves to the home venue of team 1 in the round 0 (i.e., the home venue of team 1 is the starting position of umpire 1). The solution shown in Fig. 1 would be represented as:

```
move(1,1,0).    move(1,4,3).   move(2,3,0).    move(2,2,3).
move(1,4,1).    move(1,2,4).   move(2,1,1).    move(2,3,4).
move(1,2,2).    move(1,3,5).   move(2,4,2).    move(2,1,5).
```

Obviously, in a valid solution these atoms must be such that team t actually plays a home game in round r. We first define which teams this are.

```
home_team(Home,R) :- plays(Home,Away,R).
```

We can then generate the search space by the following choice rule:

```
1 { move(X,Y,T) : home_team(Y,T) } 1 :- umpire(X), round(T).
```

Testing the hard constraints. Each game must be assigned precisely one umpire. We represent this requirement as:

```
% Each game is officiated by at most one umpire
:- move(U1,T,R), move(U2,T,R), U1 != U2.
```

```
% Each game is officiated by at least one umpire
:- home_team(T), round(R), { move(U,T,R) : umpire(U) } 0.
```

The expression {move(U,T,R) : umpire(U)} 0 holds if the set of all move(U,T,R) atoms for which round(R) holds is at most zero; in other words, if no umpires are assigned to the round R game in which team T is the home team. Given the way in which the search space is generated, it would actually be sufficient to add only of these two constraints. However, adding both of them helps the solver to solve the problem more quickly.

Next, we handle the constraint that each umpire should visit each venue. First, we define when an umpire U visits the home venue of team T:

```
been_to(U,T) :- round(R), move(U,T,R).
```

Now, the constraint is that this predicate must hold for all U and T:

```
:- umpire(U), team(T), not been_to(U,T).
```

Of the two constraints regarding repeated games with the same umpire, the constraint concerning home venues is easiest to represent (recall that limit_big contains the paramater q_1 of the instance):

```
:- move(U,T,R1), move(U,T,R2), R1 < R2, limit_big(B),
   R2 - R1 + 1 <= B.
```

To express the constraint regarding q_2, we first need to define when an umpire U officiates a game in which team T is involved (as either home or away team) in round R:

```
officiates(U,Home,R) :- move(U,Home,R).
officiates(U,Away,R) :- move(U,Home,R), plays(Home,Away,R).
```

Using this predicate, the constraint for the parameter q_1 (given by the predicate limit_small) is represented as follows:

```
:- officiates(U,T,R1), officiates(U,T,R2),
   R1 < R2, limit_small(S), R2 - R1 + 1 <= S.
```

Optimisation. The rules and constraints that we have so far suffice to produce feasible solutions. To find the optimal solution, we first define the distance D that a given umpire U travelled in round R (i.e., to go from his venue in round $R - 1$ to his venue for round R):

```
moved(U,R,D) :- umpire(U), team(T), round(R), R > 0,
                move(U,T,R), move(U,Tp,R-1), dist(T,Tp,D).
```

The following statement instructs Clasp to minimize the sum of all these distances D:

```
#minimize { D,U,R : moved(U,R,D) }.
```

5   Improving the Performance

The program that we have discussed so far correctly generates optimal solutions to the TUP. In this section, we discuss two additions to the program that may improve its computational performance.

A typical method for improving the efficiency of such a program is to introduce additional constraints to break symmetries. For the TUP, only one symmetry is known in the literature, namely the fact that the umpires are all identical. We therefore introduce the following symmetry breaking rule, which forces one particular assignment in round 0:

```
:- move(U1,T1,0), move(U2,T2,0), U1 < U2, T1 >= T2.
```

A second possibility for improving the efficiency is to choose a suitable heuristic to guide the search process. Similar to, e.g., SAT solvers, an ASP solver works by iteratively selecting some atom that is currently neither true nor false, assigning it one of these two truth values, and then propagating the effects of this assignment. A heuristic is used to decide, first of all, which atom (among those that do not yet have a truth value) should be selected and, second, whether this atom should be made true or whether it should be made false. Clasp already contains a number of built-in heuristics, but it also allows users to provide a domain-specific heuristic by defining a predicate _heuristic (a, x, w). Here, a is an atom, w is a weight and x is a special constant. If $x = $ sign, then w must be either 1 or -1. The meaning of such an atom _heuristic(a ,sign,-1) is that whenever atom a is chosen, it will be given the value false; if instead $w = 1$, then a is assigned the value true.

For the TUP, a good heuristic might be to make choices that assign an umpire to a particular game (i.e., that make a move(U,T,R) atom true). It seems likely that these are the kind of choices that will lead to the most propagation: indeed, making a single such atom true will have at least the effect of forcing all $n-1$ such atoms for the same game to be false. We therefore include the following rule:

```
_heuristic(move(U,T,R),sign, 1) :- round(R), team(T), umpire(U).
```

To complete the heuristic, it still remains to decide *which* move(U,T,R) atom to choose first. To get the most propagation, it seems reasonable to handle the different rounds in order, and since the assignment in the first round is fixed by the symmetry breaking rule, it makes sense to work in ascending order. Within a round, we greedily attempt to send each umpire to the venue with the smallest travel distance. In this way, we hope to bias the search towards more optimal solutions. This is the same strategy as followed in [14]. We implement this strategy by assigning each move(U,T,R) atom a weight of $n\delta_{max} + (\delta_{max} - \delta)$, where δ_{max} is the maximum distance between venues, δ is the distance that the umpire would have to travel to get to the home venue of team T in round R, and n is the number of rounds left after the Rth round.

```
_heuristic(move(U,T,R),level,(L-R+1) * M + (M-D) ) :-
    last(L),round(R),team(T),umpire(U),
    move(U,T1,R-1),dist(T,T1,D),max_dist(M).
```

In addition to changing the heuristic, Clasp also has numerous configuration options that determine how it precisely conducts the search for a solution. Because the space of all possible configurations is huge, a tool called Piclasp has been developed, which tries to automatically deduce an optimal configuration for a given problem using the SMAC method [12]. It does this by first experimenting with different settings on a number of training instances and then applying Machine Learning techniques to surmise an optimal configuration. This tool does not yet offer support for optimisation problems, however. Nevertheless, we have used it to determine a configuration for the problem of finding a feasible solution (i.e., one that satisfies all the hard constraints, without taking the optimisation criterium into account).

Finally, Clasp is also able to take advantage of multiple processors by running certain computations in parallel. It offers two strategies for this. The *compete* strategy launches a number of independent executions of the solver, each with different configuration parameters. The output is then simply the result produced by the best configuration. The *split* strategy essentially runs a single instance of the solver, splitting its search tree over different parallel threads. This has the advantage that the different threads actually help each other, rather than competing with each other and possible duplicating a lot of the work done by the other threads. The disadvantage, however, is that the different threads now need to communicate with—and thus wait for—each other, which may cause some of the capacity to go unused.

6   Experimental Results

In this section, we present some experiments that were conducted to determine which ways of running Clasp perform best for the TUP and how our solution compares to the state of the art. For the first point, we want to answer three questions:

- Whether the domain-specific heuristic discussed in the previous section performs better than Clasp's default heuristic (which it selects based on some properties of the problem instance);
- Whether the configuration settings learned by Piclasp perform better than the default configuration settings (which Clasp again selects based on properties of the problem instance);
- Whether using multiple processors in parallel leads to better results and, if so, which of the two possible strategies (*compete* or *split*) is better.

Experimental results are shown in Table 1. In the name of the different variants, def and dom refer to the use of, respectively, Clasp's default heurstic versus our domain-specific heuristic; 1, comp and split refer to whether only a single processor was used, or whether all available processors were used with either the *competitive* or *split* strategy; finally, π is added to the name whenever the settings learned by Piclasp where used instead of the default settings. All results in this table were computed on a Linux machine containing eight Intel(R) Core(TM) 3.2 GHz processors. The benchmarks instances come from [16] and a timeout of 10 min was used. All results in this section are given as the fraction difference with the best known solution from [19]. In other words, if x is the best value that was computed at the time-out and b is the best known value, then the tables report $\frac{x-b}{b}$. In addition, the table also reports the average fraction difference for each of the variants, the number of benchmarks for which the variant failed to produce a feasible solution with the time limit (indicated as NoSol in the table— these values are not taken into account when computing the average percentage difference), and the number of benchmarks won by this variant.

Table 1. Comparison of different variants for size $n = 14$.

Benchmark	def-comp	dom-split-π	def-1-π	def-1	dom-split	dom-1	dom-comp	dom-1-π	def-split	dom-comp-π
14-5-3	0.14	0.18	0.20	0.15	0.13	0.15	**0.13**	0.15	0.14	0.17
14-6-3	0.10	0.13	0.19	0.12	0.11	0.11	0.11	0.16	**0.09**	0.13
14-7-3	0.07	0.10	0.14	0.06	0.07	0.07	0.07	0.13	**0.06**	0.07
14A-5-3	0.17	0.21	0.23	0.18	**0.16**	0.16	0.16	0.21	0.16	0.19
14A-6-3	0.10	0.14	0.23	0.12	0.12	**0.10**	0.13	0.18	0.11	0.15
14A-7-3	**0.06**	0.12	0.20	0.11	0.09	0.09	0.07	0.11	0.08	0.12
14B-5-3	0.14	0.17	0.18	0.16	**0.13**	0.15	0.14	0.18	0.14	0.17
14B-6-3	0.12	0.14	0.15	0.13	**0.11**	0.12	0.12	0.14	0.13	0.16
14B-7-3	**0.07**	0.10	0.16	0.10	0.09	0.09	0.08	0.11	0.09	0.11
14C-5-3	0.16	0.18	0.22	0.18	**0.15**	0.17	0.16	0.21	0.19	0.18
14C-6-3	0.15	0.16	0.21	0.15	**0.10**	0.13	0.13	0.13	0.13	0.18
14C-7-3	**0.11**	0.13	NoSol	0.12	0.11	0.12	0.11	0.16	0.11	0.16
Avg	0.12	0.15	0.19	0.13	0.11	0.12	0.12	0.16	0.12	0.15
Timeouts	0	0	1	0	0	0	0	0	0	0
Wins	3	0	0	0	5	2	1	0	2	0

A first observation about these results is that the default settings consistently perform better than those learned by Piclasp, regardless of which heuristic is being used and regardless of whether parallel processing is used. This may be due to the fact that Piclasp does not support optimisation problems, or to the fact that we supplied it with relatively small training instances (because it needs to run a large number of experiments). For these reasons, Piclasp's training data may not have been representative enough for the actual problem.

Second, on a single processor, our domain-specific heuristic is almost always (in 10 of the 12 benchmarks) better than the default heuristic. In competitive multi-processor mode, the default heuristic has a slight edge (7 of the 12 benchmarks), whereas in split mode, the domain dependent heuristic has the edge (also 7 out of 12). When the default heuristic is used, competitive mode beats split mode (8 out of 12), whereas split mode always beats competitive mode for the domain-dependent heuristic. Regardless of the heuristic, using all 8 processors instead of just a single processor leads to better results *if* the appropriate multi-processor strategy is chosen (e.g., dom-1 beats dom-comp, but loses to dom-split in 9 benchmarks).

We conclude from these results that the best variants are dom-split and def-comp. In the rest of our experiments, we use these two variants, together with dom-1, the best single-processor variant. We now compare these three variants with results from [16]. This article reports on three different approaches: a Constraint Programming (CP) approach, an Integer Programming (IP) approach (both using ILOG OPL Studio 3.7 as a solver), and the special-purpose search algorithm GBNS developed by its authors. The results reported for these three approaches are taken from [16] and were performed on an Intel(R) Xeon(TM) processor which ran at the same clock speeds of 3.2 GHz as the Intel(R) Core(TM) processor we used to test our own approach.

Table 2. Comparison for $n = 14$ between our methods (timeout: 10 min) and those of [16] (timeout: 3 h).

Benchmark	def-comp	IP [16]	dom-split	GBNS [16]	CP [16]	dom-1
14-5-3	0.14	0.13	0.13	**0.08**	0.18	0.15
14-6-3	**0.10**	0.15	0.11	0.11	0.16	0.11
14-7-3	0.07	0.15	**0.07**	0.12	0.09	0.07
14A-5-3	0.17	0.12	0.16	**0.10**	0.20	0.16
14A-6-3	0.10	0.13	0.12	0.12	0.20	**0.10**
14A-7-3	**0.06**	0.17	0.09	0.11	0.09	0.09
14B-5-3	0.14	0.12	0.13	**0.08**	0.19	0.15
14B-6-3	0.12	0.16	**0.11**	0.17	0.20	0.12
14B-7-3	**0.07**	0.21	0.09	0.09	0.10	0.09
14C-5-3	0.16	**0.12**	0.15	0.14	0.20	0.17
14C-6-3	0.15	0.10	**0.10**	0.14	0.24	0.13
14C-7-3	**0.11**	0.16	0.11	0.22	0.19	0.12
Avg	0.12	0.14	0.11	0.12	0.17	0.12
Wins	4	1	3	3	0	1

Table 2 shows results for $n = 14$. Our own approaches were benchmarked with a timeout of *10 min*, while the results reported for the approaches from [16] use a timeout of *3 h*. Even taking into account the fact that our multiprocessor variants use 8 processors instead of one, this still puts our approach at a disadvantage. Nevertheless, as can be seen in Table 2, it performs quite well. All of our variants beat Trick et al.'s CP solution on all benchmarks. Against Trick et al.'s IP solution, dom-split performs best, winning 7/12 benchmarks and tying 2; our single-processor variant and def-comp also win 7 but lose the others. Against

Table 3. Comparison for $n = 16$ between our methods and those of [16] (both timeouts: 3 h).

Benchmark	def-comp	CP [16]	GBNS [16]	dom-split	IP [16]	dom-1
16-7-2	0.27	0.34	**0.11**	0.26	0.14	0.26
16-7-3	0.18	0.32	0.17	**0.14**	NoSol	0.16
16-8-2	0.16	0.24	0.11	0.18	**0.10**	0.18
16-8-4	**UNSAT**	NoSol	NoSol	NoSol	NoSol	**UNSAT**
16A-7-2	0.25	0.36	**0.09**	0.22	0.17	0.23
16A-7-3	0.17	0.18	0.17	**0.15**	NoSol	0.15
16A-8-2	0.18	0.29	**0.10**	0.16	0.11	0.19
16A-8-4	**UNSAT**	NoSol	NoSol	NoSol	NoSol	**UNSAT**
16B-7-2	0.26	0.36	**0.11**	0.24	0.14	0.25
16B-7-3	0.18	0.25	NoSol	**0.18**	NoSol	0.18
16B-8-2	0.17	0.22	**0.11**	0.14	0.12	0.17
16B-8-4	**UNSAT**	NoSol	NoSol	NoSol	NoSol	NoSol
16C-7-2	0.21	0.23	**0.09**	0.21	0.18	0.21
16C-7-3	0.14	0.20	NoSol	**0.13**	0.18	0.14
16C-8-2	**0.11**	0.21	0.12	0.12	0.15	0.12
16C-8-4	**UNSAT**	NoSol	NoSol	NoSol	NoSol	**UNSAT**
Avg	0.19	0.27	0.12	0.18	0.14	0.19
Timeouts	0	4	6	4	7	1
Wins	5	0	6	4	1	3

their special-purpose GBNS algorithm, our single-processor variant wins 6 and ties 2 benchmarks, dom-split wins 5 and tries 3, while def-comp wins 7.

To judge also the performance for somewhat bigger instances, Table 3 shows experimental results for $n = 16$. Here, our own approaches were benchmarked with the same timeout of 3 h as in [16]. An entry of UNSAT in this table means that the system was able to report that no feasible solution exists; an entry of NoSol means that the system was unable to decide whether a feasible solution exists within the time limit. Also for $n = 16$, the CP solution does not win a single benchmark against any of our approaches. Against IP, our def-comp approach wins 9 out of 16 benchmarks, while dom-split wins only 5 and loses 7. Against GBNS, def-comp wins half of the benchmarks against GBNS and loses the other half. Our dom-split approach does worse, winning only 5 benchmarks against GBNS and losing 7.

In conclusion, our approach clearly outperform the CP approach and does slightly better than IP. It is also on par with the GBNS special-purpose algorithm. However, the same cannot be said for more recent special-purpose algorithms. The currently state-of-the-art the approach of [14] is able to find the optimal solution for each of the $n = 14$ benchmarks and most of the $n = 16$

ones. The best results reported in Tables 2 and 3 above are typically around 10% worse than those of [14].

7   Conclusions

A declarative approach allows computational problems to be solved without requiring the development of special-purpose algorithms. An advantage of such an approach is that the problem specification can easily be changed, by adding or replacing certain constraints. Moreover, developments in solver technology immediately improve the performance of declarative solutions for numerous different problems. A disadvantage is that the computational performance of declarative solvers often lags behind that of special-purpose algorithms, since it may take some time to lift improvements that were made to specific algorithm for one specific problem to the level of a generic reasoning tool. However, features such as domain-specific heuristics allow domain knowledge to be used to improve the performance of the solver.

In this paper, we examined a declarative solution for the Travelling Umpire Problem, which is a challenging optimisation problem that has recently received a lot of attention. As we have shown, it can be formulated in an elegant and modular way in the declarative ASP paradigm. Because the TUP as considered in the literature is only an abstraction of a real-life problem in Major League Baseball, the flexibility to easily add or change constraints is a useful feature for this application.

Using the state-of-the-art ASP solver Clasp, we found that the performance of our approach improves on that of previous declarative solutions and is on par with the first special-purpose algorithm published for this problem. However, it cannot match the performance of current special-purpose algorithms. This suggests that these algorithms use more advanced techniques which have not yet found their way into general-purpose ASP solvers. We believe the TUP might therefore be an interesting benchmark to guide future developments in ASP solver technology.

References

1. Alviano, M., Calimeri, F., Charwat, G., Dao-Tran, M., Dodaro, C., Ianni, G., Krennwallner, T., Kronegger, M., Oetsch, J., Pfandler, A., Pührer, J., Redl, C., Ricca, F., Schneider, P., Schwengerer, M., Spendier, L.K., Wallner, J.P., Xiao, G.: The fourth answer set programming competition: preliminary report. In: Cabalar, P., Son, T.C. (eds.) LPNMR 2013. LNCS, vol. 8148, pp. 42–53. Springer, Heidelberg (2013). doi:10.1007/978-3-642-40564-8_5
2. Calimeri, F., Gebser, M., Maratea, M., Ricca, F.: Design and results of the fifth answer set programming competition. Artif. Intell. **231**, 151–181 (2016)
3. Calimeri, F., et al.: The third answer set programming competition: preliminary report of the system competition track. In: Delgrande, J.P., Faber, W. (eds.) LPNMR 2011. LNCS, vol. 6645, pp. 388–403. Springer, Heidelberg (2011). doi:10.1007/978-3-642-20895-9_46

4. de Oliveira, L., de Souza, C., Yunes, T.: Improved bounds for the traveling umpire problem: a stronger formulation and a relax-and-fix heuristic. EJOR **2**, 592–600 (2014)

5. de Oliveira, L., de Souza, C., Yunes, T.: On the complexity of the traveling umpire problem. Theor. Comput. Sci. **562**, 101–111 (2015)

6. Denecker, M., Vennekens, J., Bond, S., Gebser, M., Truszczyński, M.: The second answer set programming competition. In: Erdem, E., Lin, F., Schaub, T. (eds.) LPNMR 2009. LNCS, vol. 5753, pp. 637–654. Springer, Heidelberg (2009). doi:10. 1007/978-3-642-04238-6_75

7. Gebser, M., Harrison, A., Kaminski, R., Lifschitz, V., Schaub, T.: Abstract gringo. In: TPLP (2015)

8. Gebser, M., Kaminski, R., Kaufmann, B., Ostrowski, M., Schaub, T., Schneider, M.: Potassco: the potsdam answer set solving collection. AI Communications **24**(2), 107–124 (2011)

9. Baral, C., Brewka, G., Schlipf, J. (eds.): LPNMR 2007. LNCS (LNAI), vol. 4483. Springer, Heidelberg (2007)

10. Gebser, M., Maratea, M., Ricca, F.: The design of the sixth answer set programming competition. In: Calimeri, F., Ianni, G., Truszczynski, M. (eds.) LPNMR 2015. LNCS, vol. 9345, pp. 531–544. Springer, Cham (2015). doi:10.1007/978-3-319-23264-5_44

11. Gelfond, M., Lifschitz, V.: The stable model semantics for logic programming. In: Kowalski, R., Bowen, K. (eds.) ICLP, pp. 1070–1080. MIT Press, Cambridge (1988)

12. Hutter, F., Hoos, H.H., Leyton-Brown, K.: Sequential model-based optimization for general algorithm configuration. In: Coello, C.A.C. (ed.) LION 2011. LNCS, vol. 6683, pp. 507–523. Springer, Heidelberg (2011). doi:10.1007/978-3-642-25566-3_40

13. Toffolo, T., Van Malderen, S., Wauters, T.V., Berghe, G.: Branch-and-Price and Improved Bounds to the Traveling Umpire Problem. PATAT, York (2014)

14. Toffolo, T., Wauters, T., Van Malderen, S., Vanden Berghe, G.: Branch-and-bound with decomposition-based lower bounds for the traveling umpire problem. EJOR **3**, 932–943 (2015)

15. Trick, M.A., Yildiz, H.: Bender's cuts guided large neighborhood search for the traveling umpire problem. In: Van Hentenryck, P., Wolsey, L. (eds.) CPAIOR 2007. LNCS, vol. 4510, pp. 332–345. Springer, Heidelberg (2007). doi:10.1007/978-3-540-72397-4_24

16. Trick, M., Yildiz, H.: Benders' cuts guided large neighborhood search for the traveling umpire problem. Naval Res. Logistics (NRL) **8**, 771–781 (2011)

17. Trick, M., Yildiz, H.: Locally optimized crossover for the traveling umpire problem. EJOR **2**, 286–292 (2012)

18. Trick, M., Yildiz, H., Yunes, T.: Scheduling major league baseball umpires and the traveling umpire problem. Interfaces **3**, 232–244 (2012)

19. Wauters, T.: http://benchmark.gent.cs.kuleuven.be/tup/ Accessed 27 Oct 2016

20. Wauters, T., Van Malderen, S., Vanden Berghe, G.: Decomposition and local search based methods for the traveling umpire problem. EJOR **3**, 886–898 (2014)

21. Xue, L., Luo, Z., Lim, A.: Two exact algorithms for the traveling umpire problem. EJOR **3**, 932–943 (2015)

Student Papers

Design of a Fuzzy Logic Based Framework for Comprehensive Anomaly Detection in Real-World Energy Consumption Data

Muriel Hol[✉] and Aysenur Bilgin

Institute for Logic, Language and Computation, 1098 XG Amsterdam, Netherlands
`Muriel.Hol@student.uva.nl`, `A.Bilgin@uva.nl`

Abstract. Due to the rapid growth of energy consumption worldwide, it has become a necessity that the energy waste caused by buildings is explicated by the aid of automated systems that can identify anomalous behaviour. Comprehensible anomaly detection, however, is a challenging task considering the lack of annotated real-world data in addition to the real-world uncertainties such as changing weather conditions and varying building features. Fuzzy Logic enables modelling knowledge-based non-linear systems that can handle these uncertainties, and facilitates modelling human interpretable systems. This paper proposes a new method for annotating anomalies and a novel framework for interpretable anomaly detection in real-world gas consumption data belonging to the educational buildings of the Hogeschool van Amsterdam. The proposed architecture uses the Wang and Mendel rule learning with k-means clustering and does not require prior knowledge of the data, while preserving transparency of the model behaviour. Experiments have shown that the proposed system matches the performance of existing baselines using an artificial neural network while providing additional desired features such as transparency of the model behaviour and interpretability of the detected anomalies.

Keywords: Fuzzy logic systems · Energy consumption · Anomaly detection · Forecasting

1 Introduction

Population growth, improved indoor comfort and services, and increased time spent inside buildings cause a rapid worldwide growth of energy consumption [14]. Additionally, economic growth is causing a rise in the energy required for the buildings in the service sector such as schools, hospitals and recreational buildings. In Europe, buildings consume 40% of the entire energy, and non-residential buildings comprise the majority of this [12].

In a research conducted in 2013 [6], it was observed that buildings waste up to 30% of energy due to deficient management, which can be prevented by the utilisation of automated fault detection and diagnostics (FDD) [9]. FDD

The code is available at: https://github.com/murielhol/FuzzyEnergy.

© Springer International Publishing AG 2017
T. Bosse and B. Bredeweg (Eds.): BNAIC 2016, CCIS 765, pp. 121–136, 2017.
DOI: 10.1007/978-3-319-67468-1_9

systems detect abnormal behaviour and provide explicit information about the cause of the problem in order to enable targeted management. This detection of abnormal behaviour is described as anomaly detection [4].

The demand for better energy management in buildings through anomaly detection has resulted in various studies in the field of forecasting energy consumption [22]. The energy consumption of a building is influenced by complex features such as the buildings' materials, the users' schedule, the weather and the occupants' subjectivity regarding indoor comfort. Therefore, forecasting energy consumption requires algorithms that handle non-linearity and uncertainty.

Artificial Neural Networks (ANN) and Support Vector Machines (SVM) are the most frequently used algorithms for forecasting energy consumption as they perform well with non-linear data [7, 22]. Both methods are referred to as black-box algorithms, which means that there is no natural language explanation of the models' behaviour and hence no interpretation of mapping the input to the output of the systems [15]. As previously stated, information about the anomaly is valuable for improved energy management.

Similar to the forecasting of energy consumption, the detection of anomalies in energy data requires techniques that can handle uncertain and non-linear data and can enable transparency of the process [20]. Fuzzy Logic (FL) satisfies these requirements. After the introduction of FL by Lotfi A. Zadeh in 1965, FL has achieved success in a wide range of research fields such as control systems, image processing, industrial automation, robotics, and optimisation [17]. Zadeh states that: "Essentially, such a framework provides a natural way of dealing with problems in which the source of imprecision is the absence of sharply defined criteria of class membership rather than the presence of random variables" [21] (Zadeh, p. 339). Furthermore, FL uses linguistic terms, which facilitate understanding of the model behaviour.

The applications of FL in forecasting energy consumption and anomaly detection have yielded good results in the previous studies [18,20]. Rocha et al. [15] strengthen the argumentation for exploring FL by stating that although complex algorithms such as ANN and SVM may lead to high accuracies, non-complex transparent algorithms such as FL can provide valuable insights regarding the model behaviour.

For buildings in the service sector of The Netherlands, gas consumption comprises the largest share in energy consumption [13]. Gas consumption is severely affected by the weather, which is an uncertain feature. Therefore, the aim of this study is to tackle the uncertainties inherent to forecasting and anomaly detection in gas consumption data using FL and to provide linguistic descriptions of the identified anomalies in order to facilitate improved energy management systems. The main contributions reported in this paper are the introduction of a new method for annotating anomalies, and a novel framework for anomaly detection based on FL.

The paper is structured as follows: In Sect. 2, we present a brief background of the methods used in this research. Section 3 focuses on the related work. Section 4 details the proposed framework while Sect. 5 presents the experiments and results. In Sect. 6, we discuss our findings and lastly, we provide conclusions and future work in Sect. 7.

2   Background

This section provides a brief theoretical background on FL, Fuzzy Logic Systems, Anomaly Detection and supplementary methods we have used.

2.1   Fuzzy Logic

Zadeh describes a fuzzy set as "a class with continuum grades of membership" [21]. In other words, an element can belong to a set with a degree of membership, as opposed to classical logic, which only allows binary membership. For example, a linguistic variable, say '*Temperature*', can be modelled using two linguistic terms: '*Cold*' and '*Warm*'. These linguistic terms are characterised by membership functions, which are denoted $\mu_{Cold}(x)$ and $\mu_{Warm}(x)$, respectively. The membership functions (MFs) are used to assign membership degrees to x within the unit interval $[0, 1]$. Figure 1a illustrates the overlapping of MFs where both $\mu_{cold}(x) > 0$ and $\mu_{warm}(x) > 0$ can hold for the same value of x. However, in Fig. 1b, the Boolean sets only allow for either $\mu_{cold}(x) = 1$ or $\mu_{warm}(x) = 1$.

Modelling of MFs can be regarded as a highly problem-dependant task. Common techniques include using expert knowledge and using training data to which the MFs can be fitted. The hyper parameters to be taken into account are the number of fuzzy sets per feature, their domain (universe of discourse) and the shape of the MFs. An important design decision is the amount of overlap between the MFs. The most frequently used shapes for MFs include Triangles, Trapezoids and Gaussians.

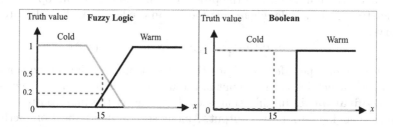

Fig. 1. (Left) a: FL example, $\mu_{Cold}(15) = 0.5$ and $\mu_{Warm}(15) = 0.2$. (Right) b: Boolean example $\mu_{Cold}(15) = 1$ and $\mu_{Warm}(15) = 0$.

2.2   Fuzzy Logic Systems

A Fuzzy Logic System (FLS), also named as Fuzzy Inference System (FIS) or Fuzzy Rule-Based System (FRBS), consists of 5 components: (1) Fuzzifier, (2) Rule Base, (3) Data Base, (4) Inference Engine and (5) Defuzzifier [16]. Figure 2 provides an overview of the FLS architecture. FLSs are universal approximators [19], capable of learning any non-linear function.

- **Fuzzification:** Fuzzification is the process of assigning a membership degree to a crisp input for a linguistic term. For example, in Fig. 1a, the fuzzification of the crisp input $x = 15$ yields 0.5 for $\mu_{cold}(15)$ and 0.2 for $\mu_{warm}(15)$, as marked on the figure.

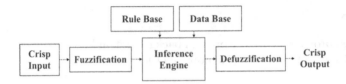

Fig. 2. Fuzzy logic system architecture

- **Rule Base:** The allowance of uncertainty reflects the human reasoning process and ambiguity of the discourse [2]. This reasoning is captured in the format of fuzzy IF-THEN Rules. The IF part of the fuzzy rule holds the antecedents and the THEN part holds the consequents. These rules can be given by an expert, or can be learnt from training examples [2]. A fuzzy rule can be formalised as:

$$\text{IF } x_1 \text{ is } \mathcal{A}_1 \text{ and ... and } x_p \text{ is } \mathcal{A}_p \text{ THEN y is } \mathcal{B} \tag{1}$$

- **Data Base:** The data base stores the linguistic variables (e.g. temperature), their linguistic terms (e.g. cold, warm) and the parameters of the MFs.
- **Inference Engine:** The inference engine infers the firing strengths of the rules for the crisp inputs, using the information from the data base and the rule base. In order to calculate the firing strength of a rule (where the antecedents are connected using the logical operator AND), the fuzzy intersection[1] (i.e. t-norm) is used [11]. The most commonly used operators for t-norm are the minimum and the product. Using the minimum operator, the firing strength of a rule where there are p antecedents can be calculated as follows:

$$FS = min(\mu_{\mathcal{A}_{x_1}}(x_1), \mu_{\mathcal{A}_{x_2}}(x_2)...., \mu_{\mathcal{A}_{x_p}}(x_p)) \tag{2}$$

In a Mamdani type inference, both the antecedents and the consequents are fuzzy sets, which leads the output of the inference engine to be a fuzzy set.
- **Defuzzification:** Defuzzification is used to convert the fuzzy output of the inference engine into a crisp output. Among several defuzzification methods, Eq. 3 formalises the centroid defuzzification where K is the number of rules, fs_i is the firing strength of the ith rule, and c_i is the centroid of the consequent of the ith rule:

$$y = \frac{\sum_{i=1}^{K} fs_i \cdot c_i}{\sum_{i=1}^{K} fs_i} \tag{3}$$

2.3 Anomaly Detection

Anomaly detection is the detection of abnormal behaviour of a data point. A frequently used method for anomaly detection is forecasting using time-series data. A data point is classified as an anomaly when the squared difference between the

---

[1] In FL, set theoretic operations (e.g. intersection, union) are defined in terms of their membership functions.

predicted and actual value exceeds a predefined threshold [4]. In their extensive survey on anomaly detection, Chandola and Kumar [4] state that: "Defining a normal region that encompasses every possible normal behaviour is very difficult. In addition, the boundary between normal and anomalous behaviour is often not precise. Thus, an anomalous observation that lies close to the boundary can actually be normal, and vice versa." (p. 3). Accordingly, it is very challenging to measure the classification performance (i.e. normal vs. anomalous) using (unannotated, real-world) data that has uncertainties with regards to what is considered as normal or anomalous.

2.4 Supplementary Methods Used in the Proposed Framework

The proposed framework uses the Wang and Mendel (WM) rule learning with k-means clustering. In this section, both methods will be presented, together with the performance measures we employed for the evaluation.

Wang and Mendel Rule Learning. Wang and Mendel provide a simple ad-hoc model for learning a fuzzy rule base from data. Their method is renowned for its "simplicity and good performance" [2] (Alcala et al. p. 11) and is referred to as the WM method. The method is based on the following steps [2]:

1. For each training example $e_l = \{x_1^l, x_2^l..., x_n^l, y^l\}$, find the sets $[\mathcal{A}_q^1...\mathcal{A}_r^n, \mathcal{C}_s]$ that e_l has the highest membership to, and create the rule

$$R_j = \text{ IF } x_1 \text{ is } A_q^1 \text{ and... and } x_n \text{ is } A_r^n \text{ THEN } y \text{ is } C_s \qquad (4)$$

   with degree $D_j = \mu_{\mathcal{A}_q}(x_1^l) \cdot .... \cdot \mu_{\mathcal{A}_r}(x_n^l) \cdot \mu_{\mathcal{C}_s}(y^l)$.
2. In order to prevent conflicting rules, if the rule base already has a rule that has the same antecedents with a different consequent, only keep the rule that has the highest degree D_j.

K-Means. K-means is a clustering algorithm that provides the centroids of the clusters in the data. In the proposed approach, the cluster centroids will be assigned as the centroids of the fuzzy sets.[2]

Performance Measures. In order to measure the forecasting performance, we employed the Mean Absolute Error (MAE) and Root Mean Squared Error (RMSE). For the anomaly detection performance, we employed the F1 score, which is a classification measure calculated with the precision and recall.

3   Related Work and Motivation

This research was mainly inspired from de Nadai and van Someren [7], who employed an ANN for the forecasting of gas consumption data. They further

---

[2] sklearn.cluster.KMeans: http://scikit-learn.org/stable/modules/clustering.html# clustering. The default settings were used in this paper.

improved their approach using an auto-regressive integrated moving average model (ARIMA) [7], which models seasonal data. The ARIMA predicted gas consumption is then added as a feature to the ANN model. Using the ARIMA features, their initial MAE of the ANN was improved from 9.52 to 7.33.

Zhao and Magoulés [22] and Ahmad et al. [1] provide an extensive review of research in energy consumption forecasting. Both studies address that SVM and ANN are the most frequently used approaches. Additionally, the studies both confirm the improvement achieved through the utilisation of ARIMA models. Neither of the studies discuss FL as an individual approach, however, they both mention several studies that obtained improved results when FL was integrated in order to form a hybrid model.

On the other hand, while considerable research is done on the prediction of energy consumption [22], few equivalent research was found on anomaly detection. One reason for this is the absence of annotated data and its consequential challenges for the performance evaluation. The lack of annotated data can be dealt with by the injection of synthetically created anomalous data [20]. However, when few and obviously anomalous synthetic anomalies are used for the evaluation, the reliability of the models become questionable. It can also be argued that the boundary problem is ignored, and hence, the models that use synthetic anomalies fail to reflect the system performance accurately, which is crucial especially in real-world scenarios.

Additionally, anomaly detection with supervised learning requires annotated data in order to confirm that the training data is free from anomalies. Moreover, the assumption that the training data is clean "constitutes a fundamental concept underlying the use of anomaly detection techniques" [20] (Wijayasekera et al. p. 1832). However, labelling the data as anomalous or normal is an expensive task that is mostly done manually. Hodge and Auston [8] provide a framework for anomaly detection. They provide supervised methods that can be used to "identify errors and remove their contaminating effect on the data set and as such to purify the data for processing" (p. 1).

An important characteristic of the anomaly detection task is the format in which the anomaly is reported [4]. Linguistic output about an anomaly allows understanding of the cause of the problem and therefore facilitates targeted action. FL naturally provides the information about the reasoning process of the system. Wijayasekera et al. successfully implemented a fuzzy anomaly detection method that avails from the linguistic reasoning of the model [20]. They propose to make the provided information about the anomaly more concise by only displaying the most important fired antecedents. They show that the firing strengths and linguistic terms of the sets can be used not only to provide information about the anomaly, but also to adjust the content of the feedback to be communicated back to the user. In our proposed method, we adopt this idea of information summarising to enable comprehensive anomaly detection.

When the Rule Base is learnt in an ad-hoc manner, it represents the behaviour of the training data [2]. Hence, when the training data represents the normal behaviour, the firing strength of a data point indicates to what extent the data point behaves normally [20]. The proposed approach exploits this feature by using the firing strengths for the anomaly classification, instead of the predicted crisp output.

4 Proposed Framework for Anomaly Detection in Gas Consumption Data

This section presents the data and the preprocessing steps, which are followed by the design of the proposed approach.

4.1 Energy Consumption Data and Feature Extraction

In this paper, we used the gas and electricity consumption rates of The Nicolaes Tulphuis (NTH) building, which belongs to HvA[3]. This data originates from

Table 1. The features used in the proposed approach

Feature	Data	Domain
Radiation	Q(t) (J/cm^2)	[0, 334]
Temperature	T(t) (0.1 C)	[−176, 331]
Wind speed	FH(t) (0.1 m/s)	[0, 220]
Humidity	U(t) (%)	[22, 100]
Temperature peak in past 5 h	$\text{argmax}_{x \in [t-6, t-1]} T(x)$	[−176, 331]
Temperature difference with previous hour	$T(t) - T(t-1)$	[−57, 63]
Gas consumption (target)	G(t)	[0, 323]
Gas consumption 1 h before	G(t−1)	[0, 323]
Gas consumption 2 h before	G(t−2)	[0, 323]
Gas peak in past 5 h	$\text{argmax}_{x \in [t-6, t-1]} G(x)$	[0, 323]
Gas peak in past 24 h	$\text{argmax}_{x \in [t-25, t-1]} G(x)$	[0, 323]
Gas sum of past 5 h	$\sum_{i=1}^{5} G(t\text{-}i)$	[0, 1098]
Gas sum of past 24 h	$\sum_{i=1}^{24} G(t\text{-}i)$	[0, 4116]
Gas mean of past 15 days	$\frac{1}{360} \sum_{i=1}^{5} G(t\text{-}361)$	[0.41, 132.11]
Day of the week		[0, 6]
Next day of the week		[0, 6]
Hour of the day		[1, 24]
Day of the year		[0, 366]
Electricity consumption (in kWh)	E(t)	[0, 53.11]
Electricity consumption peak in past 5 h	$\text{argmax}_{x \in [t-6, t-1]} T(x)$	[0, 53.33]
STD residuals for trend = day	resDay(t)	[−116.24, 109.28]
STD residuals for trend = year	resYear(t)	[−117.85, 164.19]

---

[3] Hogeschool van Amsterdam: http://www.hva.nl/over-de-hva/locaties/locaties.html.

the research in de Nadai and van Someren [7] and consists of 52608 hourly gas consumption timestamps ranging from the first day of 2008 until the first day of 2014. There were 2 missing dates, 29 missing gas values and 6 missing electricity values, which were linearly interpolated.

We adopted the features that are used by de Nadai and van Someren [7] in order to enable a fair comparison with their results. Moreover, the same features were also adopted by Lodewegen [10], who also studied the performance of ANN on this data set. Both studies will serve as reference points for the results presented in this paper. In total, there are 22 features (see Table 1), which can be categorised under the following 4 major interests:

1. **Weather data:** Hourly weather data was obtained from the KNMI website[4]. The Schiphol weather station is the closest one to the NTH building (± 20 km).
2. **Energy consumption time series:** Each data row has extensive information about the historical gas and electricity consumption, such as the gas consumption in the previous hour or the peak in consumption during the last 24 h.
3. **Time stamps:** Information about the moment when the data was recorded is provided in terms of day of the week and the hour of the day. The next day of the week is also included as a feature, since schools tend to warm up the buildings for Monday in advance [7]. National holidays were labelled as Sundays.
4. **Seasonal Trend Decomposition:** Gas consumption is strongly related to seasonal trends in for example days and years. In order to allow the model to understand the behaviour of the consumption in a disconnected manner from its seasonal trend, the LOESS method [5] for Seasonal Trend Decomposition (STD) is adopted [7]. By using the LOESS method, the gas consumption is decomposed into seasonality, trend and residual. As emphasised by [7], the residual is very informative for the prediction of gas consumption. Figure 3 shows the decomposing for the days and years from the beginning of 2008 until the end of 2013.

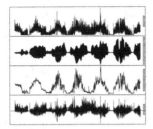 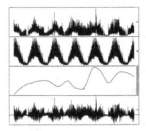

Fig. 3. Seasonal trend decomposition of the gas consumption. From top to bottom: the data, the seasonal component, the trend and the residuals. (Left) a: STD for trend day (frequency is 24). (Right) b: STD for trend year (frequency is 24*365.5).

---

[4] Koninklijk Nederlands Meteorologisch Instituut: http://projects.knmi.nl/klimatolog ie/uurgegevens/selectie.cgi.

In the proposed FLS, each feature is a linguistic variable such as *Temperature*, and the feature values are linguistic terms such as *very low, high, somewhat high* etc.

4.2 Pre-processing of the Data and Annotation of Anomalies

In order to detect anomalies, the data needs to be labelled as normal or anomalous. Since the original data is not annotated, we propose to synthetically label the data using the 3 sigma rule and the Mahalanobis distance. The 3 sigma rule assumes that the probability of a data point lying in the region ± 3 times the standard deviation (σ) of the mean (μ) is 0.997. This is formalised as follows: $P(\mu - 3\sigma \leq X \geq \mu + 3\sigma) \approx 0.997$.

Since the data is multivariate, the Mahalanobis distance measure is used to calculate the distance to the mean vector [8]. Mahalanobis distance measure is formalised as follows: $MD = \sqrt{(\boldsymbol{x} - \boldsymbol{\mu})^T \Sigma^{-1} (\boldsymbol{x} - \boldsymbol{\mu})}$. Here, Σ is the covariance matrix. The covariance matrix is a matrix that holds the variance for each feature combination. Therefore, the dependencies between all features are taken into account. In the rest of the paper, the inliers will be referred to as 'clean data'. We employ k-fold cross validation where $k = 5$, hence, the data will be split into 5-folds of 80% for training, and 20% for testing.

4.3 Architecture of the Proposed Framework

The proposed framework is illustrated in Fig. 4 and is based on the following steps:

1. Data Base construction: We used k-means clustering to find an optimal centroid location for each MF. The linguistic terms that describe the MFs depend on the number of clusters and are listed below:
    - For 11 clusters: Extremely low, very low, low, somewhat low, a bit low, medium, a bit high, somewhat high, high, very high, extremely high
    - For 9 clusters: Extremely low, very low, low, a bit low, medium, a bit high, high, very high, extremely high
    - For 7 clusters: Extremely low, very low, low, medium, high, very high, extremely high
2. Rule Base construction: We adopted the ad-hoc rule learning method constructed by Wang and Mendel [19] as presented in Sect. 2.4.
3. Anomaly classification threshold: A threshold was constructed based on the firing strengths of the anomalies using the part of the data between 2008 and 2012. For a chosen anomalous data point, we calculated the mean firing strength using the entire rule base. We repeated this for all the anomalous points that are obtained from the 3 sigma rule. We then calculated the mean firing strength of all 480 anomalous points and used this as a threshold ($\overline{f}$). A data point is classified as an anomaly when its mean firing strength over all the rules is lower than $\overline{f}$.
4. Other FLS design parameters: We used Mamdani inference, with minimum t-norm and centroid defuzzification.

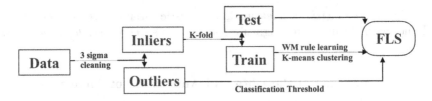

Fig. 4. The architecture of the proposed framework

5   Experiments and Results

In this section, we present several experiments on the following: (1) tuning of the FLS, (2) validation of the forecasting performance and (3) validation of the anomaly detection performance. Furthermore, we illustrate the capabilities of FL based approach by describing a couple of anomalies.

5.1   Tuning Experiments and Results

The tuning of the FLS consists of several cycles of parameter tuning until the lowest forecasting error is obtained through trial and error. Once the configuration is found, then the final rule base and data base are stored. The tuning tests were done on the data from the year 2013, in order to reflect real life performance in which the system is trained on historical data and tested on the forthcoming.

The system is tuned for (a) the MF type (i.e. Triangle, Trapezoidal or Gaussian) (b) the number of MFs per feature, and (c) the domains of the MFs. For tuning the MF type, we fixed the number of MFs to be 9 per feature. The initial variance of the Gaussian MF was 0.1 (in a normalised domain of 0 to 1) and the initial base of the Triangle/Trapezoidal MF was 0.2 (in a normalised domain of 0 to 1). Gaussian MFs gave the best performance.

Table 2. The number of clusters (k) per feature (after tuning)

k	Features
11	Radiation, Temperature, Wind speed, Humidity, Gas consumption (target), Gas sum of past 24 h, Hour of the day, Day of the year
9	Temperature peak in past 5 h, Gas peak in past 24 h, Gas sum of past 5 h, Gas mean of past 15 days, Electricity (in kWh), Electricity peak in past 5 h, STD residuals for trend = day, STD residuals for trend = year
7	Gas consumption 1 h before, Gas consumption 2 h before, Gas peak in past 5 h, Day of the week, Day of the year

For tuning the number of MFs per feature (i.e. linguistic variable), we have chosen to place the mean of the Gaussian MFs to be the cluster centre obtained from k-means clustering that was conducted on the mean target gas values of each training example. Formally, for each unique feature value x, we calculate

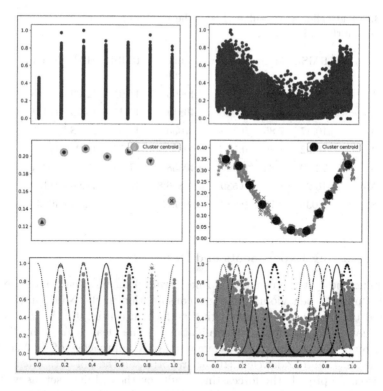

Fig. 5. The process of tuning the features: (Left) day of the week and (Right) day of the year. From top to bottom: (1) The data scaled between 0 and 1, (2) the cluster centroids according to k-means on the mean target values (3) fitting the MFs using their mean at the centre of the clusters, and a variance of 0.07.

the mean gas consumption and collect these in a vector that is as large as the number of unique feature values $\overline{y}$. E.g. if the data contains 3 examples which 'Temperature' value is 5, and they have a 'Gas consumption' value of 1, 2 and 3, respectively, then the $\overline{y}_{temperature=5} = 2$. Then, we perform k-means on x and $\overline{y}$ with a predefined $k \in 5, 7, 9, 11$. These values were chosen by trial and error until the MAE did not decrease. In order to prevent over-fitting, the maximum k was set to 11. For some features, the number of clusters were manually assigned based on the authors' visual observations on data plots. The number of clusters per feature is displayed in Table 2.

After having learnt the cluster centres, which are to be used as the mean of the Gaussian MFs, the variance of the Gaussian MFs was tuned using the following values for variance $\sigma \in \{0.06, 0.07, 0.08, 0.09, 0.1, 0.2\}$. Configuration using $\sigma = 0.07$ gave the lowest forecasting error (i.e. MAE). The values were chosen by trial and error. Figure 5 demonstrates the process of tuning for two different features.

Table 3. 5-fold validation results: (Left) The results on the clean data set. (Right) The results on the full data set

Fold	RMSE	MAE	F1-score	Number of rules	RMSE	MAE	Number of rules
1	10.11	7.76	0.53	40906	14.01	11.09	41145
2	10.15	7.8	0.56	40814	17.47	13.39	41431
3	10.47	7.85	0.53	41035	14.32	10.78	41406
4	10.92	8.40	0.58	40821	14.72	11.28	41249
5	12.90	9.84	0.40	40858	17.14	13.28	40291
Mean	10.91	8.33	0.539	40886.8	15.53	11.96	41104
STDEV	1.035	0.788	0.033	80.982	1.624	1.265	469.5

5.2   Forecasting Experiments and Results

By training the system (using WM rule learning) on the clean data, we modelled the 'normal' behaviour of the gas consumption in the NTH Building. We evaluated the model for the forecasting performance using the RMSE and the MAE. The results of 5-fold cross validation on the cleaned data are listed in Table 3. Since the results of the referenced ANN approaches of de Nadai and van Someren [7] and Lodewegen [10] are reported on the full data set, for a fair comparison, we provide the forecasting results on the full data set, as well. The results on the cleaned data set yield an RMSE of 10.91 and an MAE of 8.33. For the full data set, the MAE is 15.53 and the RMSE is 11.96.

We performed statistical T-tests in order to compare the results of the FLS and the ANN approaches. We observed that the MAE and RMSE of the proposed FL based framework are significantly higher than the ANN approach proposed by de Nadai and van Someren [7], however, not significantly different when compared to the approach of Lodewegen [10]. Table 4 shows the comparison results of all three approaches.

Table 4. Comparison with ANN approaches [7,10] on the full data set

Approach	RMSE	MAE	Configuration
de Nadai and van Someren [7]	N/A	9.52	15 epochs, 60 neurons, 1 hidden layer
Lodewegen [10]	15.58	12.01	150 epochs, 150 neurons, 1 hidden layer
Proposed FL Approach	15.54	12.07	Gaussian mf, 7–11 sets per feature, 41104 rules

5.3   Anomaly Detection Experiments and Results

In order to evaluate the performance of the anomaly detection, we annotated the data as described in Sect. 4.2. Our approach for anomaly detection relies on the significant difference between the firing strengths of the inliers and the outliers (i.e. anomalous data). For the purpose of illustrating that the difference is indeed significant, we plotted the firing strengths of both categories in Fig. 6. The mean firing strength of the inliers is 5.33e–05 (Fig. 6a) whereas the mean firing strength of the outliers is 5.79e–07 (Fig. 6b). Hence, it can be deduced that the difference between the mean firing strength of inliers and the outliers is significant. The performance on anomaly detection is validated using the same cross-validation folds as we used for the forecasting. The F1-score results for each fold are listed in Table 3.

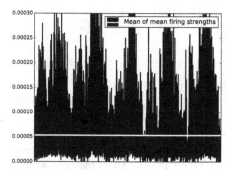 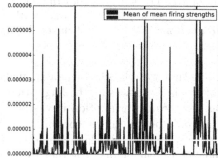

Fig. 6. (Left) a: the average firing strength of the 52128 inliers with a mean average of 5.33e–05. (Right) b: The firing strengths of the 480 outliers with a mean average of and 5.79e–07.

For further evaluation and comparison purposes, a baseline for anomaly detection has been set. A classifier that classifies an anomaly with 50% chance has been chosen to be the baseline. For the baseline system, we obtained an F1-score of 0.074. For the proposed framework, we observed the average F1-score over 5-folds to be 0.539, which is far above the baseline. However, the imbalance of the classes within data, which consists of 52129 inliers and 481 outliers, has an influence on the F1-score. Therefore, the confusion matrix with the True Positives (TP), False Positives (FP), True Negatives (TN) and False Negatives (FN) is presented in Fig. 7 in order to provide full information on the classification performance.

5.4   Linguistic Description of Anomalies

In order to exploit the advantages of the proposed FL based framework and to allow for interpretability of the anomalies, we adopted the method used by Wijayasekera et al. [20]. Accordingly, the anomalies can be described using the fuzzy rules. Furthermore, the most influential features can be detected by looking at the least fired antecedents of the rule. To give an example, one of the data

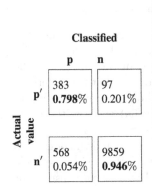

Classified

	p	n
p′	383 **0.798**%	97 0.201%
n′	568 0.054%	9859 **0.946**%

Actual value

HUMIDITY IS *A BIT HIGH* ENERGY CONSUMPTION PEAK OF PAST 5 HOURS (KWH) IS *VERY LOW* GAS CONSUMPTION PEAK IN THE PAST 5 HOURS IS *HIGH* GAS CONSUMPTION SUM IN THE PAST 5 HOURS *LOW* TEMPERATURE IS *A BIT HIGH* TEMPERATURE DIFFERENCE WITH THE PAST HOURS IS *SOMEWHAT LOW* RADIATION IS *EXTREMELY LOW*
DAY OF THE YEAR IS *EXTREMELY LOW* ENERGY CONSUMPTION PEAK OF PAST 5 HOURS (KWH) IS *LOW* GAS CONSUMPTION PEAK IN THE PAST 5 HOURS IS *LOW* GAS CONSUMPTION ONE HOUR BEFORE IS *MEDIUM* HOUR OF THE DAY IS *A BIT LOW* DEVIATION OF THE DAILY GAS CONSUMPTION TREND IS *EXTREMELY HIGH* GAS CONSUMPTION PEAK IN THE PAST 24 HOURS IS *LOW*

Fig. 7. Confusion matrix: left upper: TP, left bottom: FN, right upper: FP, right bottom: TN.

Fig. 8. Two examples of linguistic description of anomalies (Top) a: conflicting weather and (Bottom) b: 1st of January (holiday)

points that was classified as an anomaly is displayed with its 7 most influential features in Fig. 8a using the linguistic terms provided in Sect. 4.3. This example has low mean consumption in the past 5 h, however high peak in the past 5 h. This could be due to the combination of extremely low radiation yet a bit high temperature, and a bit high humidity (e.g. rain on a warm day may have caused anomalous behaviour). Another example is given in Fig. 8b. This was a data point recorded on the 1st of January, which was a holiday, and the seasonal residual for the year trend was extremely high. The linguistic descriptions can be very useful to the building managers, who wish to understand the cause of the anomalies and take targeted action.

6   Discussion

Although the forecasting results validate the performance of the FLS, it is important to note that the STD residuals at a time unit are used to classify a data point at that same time unit. Hence, if the system is only to be utilised for forecasting, then it would be better to adjust the features to include solely historical values (e.g. residuals at t−1). However, since an anomaly can be reported one time unit later, these are realistic features for anomaly detection.

With 22 features and more than 41700 training examples, the WM method for rule learning leads to a very large rule base. A rule reduction method, which is the Cooperative Rule Approach of Casillas et al. [3], was taken into consideration. However, it was found that the complexity of this method with N rules and k values for the target variable is at approximately $(N/k)^k$. Hence, for this research this would be $(10422/9)^9$, and therefore computationally infeasible. With regards to the parameter tuning, the order in which the parameters were tuned was arbitrarily chosen, which could therefore have lead to a local optimum.

7   Conclusions and Future Work

In this paper, we proposed a FL based framework for comprehensive anomaly detection in gas consumption data. The WM method and k-means clustering algorithm were combined into a supervised method for learning a fuzzy rule base, which represents the normal gas consumption behaviour of the NTH building. Furthermore, we introduced a new method for annotating anomalies using 3 sigma rule and Mahalanobis distance.

Regarding the forecasting of energy consumption, the performance of the proposed framework meets one of the existing approaches that was based on an ANN, however, is outperformed by the other. For the anomaly detection performance, we employed two techniques for evaluation: (1) we visually validated the efficiency of the proposed framework and (2) we compared the performance of the proposed framework with a baseline. We showed that the FL based approach is capable of detecting anomalies in gas consumption data far above the baseline. Furthermore, we exploited the advantages of FL based approach and demonstrated that the causes of the anomalies can be linguistically described.

For additional future work, we will investigate rule reduction techniques on high dimensional data. These techniques use optimisation methods such as simulated annealing. Finally, the informative capabilities of the proposed FL approach would be especially beneficial when applied on a data set that includes sensor data.

Acknowledgment. This study is partially supported by the Marie Curie Initial Training Network (ITN) ESSENCE, grant agreement no. 607062.

References

1. Ahmad, A.S., Hassan, M.Y., Abdullah, M.P., Rahman, H.A., Hussin, F., Abdullah, H., Saidur, R.: A review on applications of ANN and SVM for building electrical energy consumption forecasting. Renew. Sustainable Energy Rev. **33**, 102–109 (2014). doi:10.1016/j.rser.2014.01.069
2. Alcala, R., Casillas, J., Cordón, O., Herrera, F., Zwir, S.J.T.: Techniques for learning and tuning fuzzy rule-based systems for linguistic modeling and their application. In: Leondes, C. (ed.) Knowledge Engineering: Systems, Techniques and Applications, pp. 889–941. Academic Press, San Diego (1999)
3. Casillas, J., Cordón, O., Herrera, F.: Improving the Wang and Mendel's fuzzy rule learning method by inducing cooperation among rules. In: Proceedings of the 8th Information Processing and Management of Uncertainty in Knowledge-Based Systems Conference, pp. 1682–1688 (2000)
4. Chandola, V., Kumar, V.: Anomaly detection: a survey. ACM Comput. Surv. **51**, 58 (2009). doi:10.1145/1541880.1541882
5. Cleveland, R.B., Cleveland, W.S., McRae, J.E., Terpenning, I.: STL: a seasonal-Trend decomposition procedure based on loess. J. Official Statist. **6**, 3–73 (1990)
6. Colmenar-Santos, A., de Lober, L.N.T., Borge-Diez, D., Castro-Gil, M.: Solutions to reduce energy consumption in the management of large buildings. Energy Build. **56**, 66–77 (2013). doi:10.1016/j.enbuild.2012.10.004

7. De Nadai, M., van Someren, M.: Short-term anomaly detection in gas consumption through arima and artificial neural network forecast. In: Proceedings of 2015 IEEE Workshop on Environmental, Energy, and Structural Monitoring Systems (EESMS 2015), pp. 250–255. IEEE, New York (2015). doi:10.1109/EESMS.2015.7175886
8. Hodge, V.J., Auston, J.: A survey of outlier detection methodologies. Artif. Intell. Rev. **22**, 85–126 (2004). doi:10.1007/s10462-004-4304-y. Kluwer Academic Publishers, Department of Computer Science, University of York
9. Katipamula, S., Brambley, M.R.: Review article: methods for fault detection, diagnostics, and prognostics for building systems: a review, Part I. HVAC & R. Res. **11**, 3–25 (2005)
10. Lodewegen, J.: Saving energy in buildings using an Artificial Neural Network for outlier detection. B.Sc Thesis. University of Amsterdam (2015). https://esc.fnwi.uva.nl/thesis/centraal/files/f1992667963.pdf
11. Mendel, J.M.: Uncertain Rule-Based Fuzzy Logic Systems: Introduction and New Direction. Prentice Hall, Upper Saddle River (2001). ISBN: 978-0130409690
12. ODYSEE: an analysis based on the ODYSSEE and MURE databases. http://www.odyssee-mure.eu/publications/br/energy-efficiency-trends-policies-buildings.pdf
13. ODYSEE. Energy efficiency trends and policies in the Netherlands (2015). http://www.odyssee-mure.eu/publications/national-reports/energy-efficiency-netherlands.pdf
14. Perez-Lombard, L., Ortiz, J., Pout, C.: A review on buildings energy consumption information. Energy Build. **40**, 394–398 (2008). doi:10.1016/j.enbuild.2007.03.007. Elsevier Inc
15. Rocha, A., Papa, J.P., Meira, L.A.A.: How far you can get using machine learning black-boxes. In: Conference on Graphics, Patterns and Images, vol. 16, pp. 1530–01834. IEEE (2010). doi:10.1109/SIBGRAPI.2010.34
16. Sanz, F., Ramrez, J., Correa, R.: Fuzzy inference systems applied to the analysis of vibrations in electrical machines. Fuzzy Inference Syst. Theory Appl. (2012). doi:10.5772/37448
17. Singh, H., Gupta, M.M., Meitzler, T., Hou, Z., Garg, K.K., Solo, A.M.G., Zadeh, L.A.: Real-Life Applications of Fuzzy Logic, Advances in Fuzzy Systems. Hindawi Publishing Corporation (2013). doi:10.1155/2013/581879
18. Suganthi, L., Iniyan, S., Samuel, A.A.: Applications of fuzzy logic in renewable energy systems: a review. Renew. Sustain. Energy Rev. **48**, 585–607 (2015). doi:10.1016/j.rser.2015.04.037
19. Wang, L.X., Mendel, J.M.: Generating fuzzy rules by learning from examples. IEE Trans. Syst. Syst. Man Cybern. **22**, 1414–1427 (1992)
20. Wijayasekera, D., Linda, O., Manic, M., Rieger, C.: Mining building energy management system data using fuzzy anomaly detection and linguistic descriptions. IEEE Trans. Industr. Inform. **10**, 1829–1839 (2014). doi:10.1109/TII.2014.2328291
21. Zadeh, L.A.: Fuzzy sets. Inf. Control **9**, 338–353 (1965)
22. Zhao, H., Magoules, F.: A review on the prediction of building energy consumption. Renew. Sustain. Energy Rev. **16**, 3586–3592 (2012). doi:10.1016/j.rser.2012.02.049

Fostering Relatedness Between Children and Virtual Agents Through Reciprocal Self-disclosure

Franziska Burger[1(✉)], Joost Broekens[1], and Mark A. Neerincx[1,2]

[1] Delft University of Technology, Delft, Netherlands
{f.v.burger,d.j.broekens,m.a.neerincx}@tudelft.nl
[2] TNO, Soesterberg, Netherlands

Abstract. A key challenge in developing companion agents for children is keeping them interested after novelty effects wear off. Self Determination Theory posits that motivation is sustained if the human feels related to another human. According to Social Penetration Theory, relatedness can be established through the reciprocal disclosure of information about the self. Inspired by these social psychology theories, we developed a disclosure dialog module to study the self-disclosing behavior of children in response to that of a virtual agent. The module was integrated into a mobile application with avatar presence for diabetic children and subsequently used by 11 children in an exploratory field study over the course of approximately two weeks at home. The number of disclosures that children made to the avatar during the study indicated the relatedness they felt towards the agent at the end of the study. While all children showed a decline in their usage over time, more related children used the application more, and more consistently than less related children. Avatar disclosures of lower intimacy were reciprocated more than avatar disclosures of higher intimacy. Girls reciprocated disclosures more frequently. No relationship was found between the intimacy level of agent disclosures and child disclosures. Particularly the last finding contradicts prior child-peer interaction research and should therefore be further examined in confirmatory research.

1 Introduction

Type 1 diabetes mellitus (T1DM) is an autoimmune disease of the pancreas that requires manual control of blood glucose levels. Strict adherence to a medical regimen is crucial to prevent many of the health risks associated with this chronic disease. T1DM accompanies diagnosed children and adolescents through various physical and mental stages of development. We develop a **P**ersonal **A**ssistant for a healthy **L**ifestyle (PAL[1]) with the aim of increasing the self-management skills of diabetic children (ages 7–14) by supporting them, their caregivers, and healthcare professionals in sharing responsibility. The PAL robot and its mobile avatar

---

[1] http://www.pal4u.eu/.

© Springer International Publishing AG 2017
T. Bosse and B. Bredeweg (Eds.): BNAIC 2016, CCIS 765, pp. 137–154, 2017.
DOI: 10.1007/978-3-319-67468-1_10

are companion agents intended to function as a pal for the children, helping them to accomplish their diabetes-related goals through person- and time-adaptive, engaging interactions.

Since children are susceptible to novelty effects, and the PAL solution can only be effective when children continue to engage with the system, ways of sustaining their motivation are highly desirable. We are interested in exploring the possibilities and limitations of creating a bond between diabetic child (8–12 years) and the PAL virtual companion agent through self-disclosure with the goal of increasing the motivational capacity of the agent. To this end, we conducted a two-week exploratory field study in which 11 children were given the opportunity to interact with the self-disclosing mobile PAL avatar by either disclosing as well or by simply listening.

2   Theoretical Foundation

Companion agents are developed for long-term use. A key challenge in the field is thus the maintenance of motivation when novelty effects wear off.

Social relationships often play a large motivational role in our behaviors. According to Self Determination Theory (SDT), successful establishment of a social bond between human and agent leads to sustained motivation both to interact with the agent and to engage in activities that the agent proposes. SDT [7] argues that the basic psychological needs for *autonomy*, *competence*, and *relatedness* must be satisfied by the social environment for humans to feel motivated to attempt a task. Relatedness here refers to the feeling that one is accepted and cherished by another individual or community. It comes into play when the intrinsic motivation to engage in an activity is low. More simply put: if we like or want to be liked by someone, we feel more inclined to do what they suggest, even if we are not too fond of the activity itself.

An important mechanism by which such a bond could be established is described by Social Penetration Theory (SPT) [1]. It proposes a directional development of interpersonal relationships whereby the involved individuals first share and explore each others personalities at a superficial level before disclosing more intimate information. Disclosing proceeds along two dimensions: breadth and depth, with *breadth* describing the number of different topics that are disclosed about and *depth* describing the personal value these topics have. Finally, an important determinant of self-disclosure is reciprocity. This describes the tendency to self-disclose as a result of being disclosed to. Reciprocal disclosures in successfully progressing relationships are usually on a similar level of intimacy.

3   Related Literature

One of the key interests in human-human self-disclosure research has been the close link between disclosure and liking. Specifically, three persistent disclosure-liking relations have been identified [6]: (a) the more someone intimately discloses to us, the more we like that person, (b) the more we like someone at the outset of the interaction, the more we will disclose, and (c) the more intimately we disclose to someone, the more we like that person.

When children were asked what a friend is and what differentiates a friend from a non-friend, children older than nine indicated that friends take an interest in each others problems and care for their friends' emotional well-being. Additionally, cooperation and the insight that each child should contribute equally to the interaction can be expected in this age group [15]. In line with this, in a study conducted in the United States, it was found that 6th grade children's liking of another child was influenced by that child's ability to match the intimacy level of a disclosure while that of 4th graders was not [13].

Support for the disclosure-liking effect has also been found in the domains of human-computer (HCI) and child-robot (cHRI) interaction. In [11], a computer first disclosed some information about itself before asking the user (all university undergraduates) an interview question. As hypothesized, interviewees shared more intimate information with the computer that told personal information about itself but only if this personal information would gradually increase in intimacy throughout the interview. However, the liking for the computer only depended on the sharing of personal information and was not influenced by the intimacy strategy. When a robot was used to elicit self-disclosures from children (aged 10–14), those who were prompted to disclose to the robot described the robot significantly more often as a *friend* than children in the control condition [10]. In [9], a two-month study was conducted in an elementary school with a relational robot capable of identifying children (aged 10–11) and calling them by name, showing more varied behavior with time, and disclosing personal information as a function of a child's interaction time. It was found that children's desire to be friends with the robot at the end of the study was positively correlated with the interaction time.

To the best of our knowledge, there has been no empirical investigation of whether and how the sharing of disclosures between user and system contributes to sustaining user motivation over longer periods of time. The here described research presents a first step in closing this knowledge gap. We developed the initial prototype of a dyadic disclosure dialog module (3DM, Sect. 4) to gain insights into how and how readily diabetic children respond to self-disclosures of an embodied conversation agent (ECA) and to learn about the possibilities of sustaining children's motivation in this way. We were particularly interested in the following research questions, summarized by the relationship *disclosure → relatedness → motivation*:

RQ1 Can the relatedness that the child feels towards the avatar be predicted from (1) the amount of disclosures that the child heard from the avatar, (2) the amount of disclosures the child made to the avatar, and (3) the relatedness the child felt at the outset of the study?

RQ2 Is relatedness a good indicator for children's motivation to use the application?

Furthermore, we were interested in learning about how children respond to a self-disclosing avatar. Studies on self-disclosure reciprocity in child-child interaction have been conducted mainly in North America several decades ago (compare [5,12,13]). It was therefore uncertain whether insights transfer to today's children in Europe or to child-robot interaction. We thus also pursued the following research questions:

RQ3 How do children respond to the disclosures of the avatar?
(a) Is there a relationship between the avatar intimacy and children's responsiveness?
(b) Do children match the intimacy level of the avatar disclosure when they respond?
(c) What role do age and gender of the children play in how children respond to the avatar?

A situated approach was taken by integrating the module (described in the following section) into a mobile application for diabetic children to be used in an uncontrolled environment for a period of two weeks. In so doing, we found that while children did not match the intimacy of disclosures from the ECA, those children who replied more actively to the disclosures also felt more related to the avatar. Furthermore, children were more likely to reciprocate a disclosure when it was of lower intimacy or when the child was a girl.

4   Dyadic Disclosure Dialog Module

The first prototype of the dyadic disclosure dialog module (3DM) was developed. While it is the ultimate goal of the module to manage the sharing of personal information between agent and child in an adaptive and engaging manner, the first prototype, developed for this research project, only served the purpose of exploring the disclosure behavior of the children when interacting with a self-disclosing ECA. For this, content that the ECA could disclose was needed. Below we therefore briefly touch on the steps taken to develop the disclosure database and its structure. This is followed by a description of how the module functions and how it is integrated into application of the PAL project.

4.1   Content of 3DM

To design suitable disclosures for the ECA, three preliminary steps had to be taken. First, a personality for the avatar was crafted by determining sensible traits for the given domain (e.g. the ECA should be conscientious because this is important in diabetes self-management and we would like the ECA to provide positive examples of self-discipline for the children). The Murphy-Meisgeier Type Indicator for Children[2] was employed for finding a suitable type to integrate these initial traits into one coherent personality. Second, a background story was written for the robot by the lead researcher from which consistent disclosures at various intimacy levels could be derived. Here, the goal was to obtain a story that is both in line with the fact that robots are not human and in line with a character that children can embrace[3]. Third, a scaling method (rating scale) to

---

[2] https://www.capt.org/.
[3] http://latd.tv/Latitude-Robots-at-School-Findings.pdf.

design agent disclosure statements at various intimacy levels and to assess the depth of children's disclosures was developed [4].

The current disclosure database consists of approximately 150 English disclosures for the avatar at all four intimacy levels of the rating scale. These were written by the lead author taking into consideration the personality of the robot and its background story. In designing the rating scale, a selection of the statements were evaluated as a set by ten participants with regard to believability and consistency (*mean* = 4.3 on 5-point Likert-scale) with the designed personality and background story. The disclosure statements are organized into the four categories *food, school, social,* and *sports*. They also have valence labels so as to be matched to the child's affective state if available. Since two of the partner hospitals of the PAL project are in the Netherlands and the study was carried out with Dutch children, all disclosures have Dutch translations. Translations were first made by the lead author, a non-native speaker. They were then double checked and edited by two native Dutch speakers, including the second author. Finally, the Dutch translations (as these were to be used in the experiment) were investigated for their age appropriateness by the developmental psychologist, also a native Dutch speaker, involved in the PAL project. The disclosures are stored as instances of the Disclosure class—a class in the associated ontology described below.

4.2   Functionality of 3DM

Within the PAL-project, knowledge is represented in ontologies. Specifically for the module, a small ontology was therefore made. There are three main classes in the ontology for 3DM: Disclosure, Prompt, and Closer. These correspond to the three types of statements that 3DM relies on. All disclosures have the parameters intimacy level, valence, and topic. Agent disclosures additionally have an associated prompt. Prompts are said by the agent to elicit a disclosure from the child. Closers are used to end the off-activity chat and return to the main activity: a *positive* closer is said when the child chooses to disclose something, a *negative* closer is said otherwise[4]. Since the module is not yet capable of comprehending a child's disclosure, closers are very general statements that make no reference to the disclosure content. The ontology is specified in RDF[5].

The flow of the disclosure module follows a loop. From the perspective of the user this proceeds as illustrated in Fig. 1. While inactive, 3DM waits for a trigger event from the interface. When it receives this, it selects a disclosure and sends it with a gesture to the avatar for rendering. Upon execution, it follows up with the prompt. The interface then provides a pop-up asking the child whether it would like to respond. If the child chooses not to (*passive interaction*), a

---

[4] It is important to note that *positive* and *negative* here are not synonymous with rewarding or punishing the child. An example for a positive discloser can be found in the example dialog. An example for a negative closer is "That's alright. Maybe next time! In any case, thanks for listening.".

[5] https://www.w3.org/.

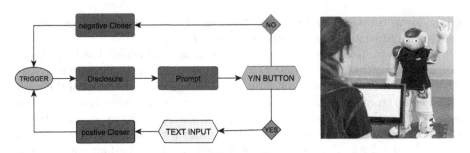

Fig. 1. *Left.* Illustration of the 3DM functionality. Interface actions are hexagonal, agent actions are rectangular, and child actions are diamonds. The trigger event (opening of the diabetes diary) has a circular shape. *Right.* A diabetic child interacts with the PAL robot. Photo courtesy of Rifca Peters.

negative closer command is sent to the avatar. If the child wants to respond (*active interaction*), it can do this in a second pop-up that allows it to type some text. Once the module has received the text, it sends a positive closer command to the avatar. It then simply waits for the next trigger event. In the first prototype, the trigger event was chosen to be the opening of the diabetes diary area of the app. Both closer sentences and prompt sentences contain a placeholder for using the name of the child. It is randomly decided whether to use the name in the prompt, in the closer, or not at all.

An example dialog of the agent (**A**) with a fictional child (**C**) called Maria may look like this:

C selects diary feature of application to access the diabetes diary. Before diary opens:
A(disclosure): "I also go to school! Together with all the other robots at the hospital. Our teachers are doctors and nurses."
A(prompt)   : "Enough about me! Tell me something interesting about yourself!"
Interface   : *Would you like to tell NAO something?* yes/no
C(selecting) : yes
Interface   : *Please provide your response below.* text input field
C(typing)   : "I had a lot of fun at school today. We played hide and seek during the break. No one found me!"
A(p. closer) : "Thanks for sharing that with me, Maria!"
Diabetes diary opens

5 Method

To investigate the relationship *disclosure → relatedness → motivation*, a two-week, exploratory field study was conducted.

5.1 Participants

Participants in the study were 11 diabetic children between the ages of 8 and 12 ($Mean_{age} = 9.91\,years$, $SD_{age} = 1.08\,years$, 6 girls). All participants had previously participated in an evaluation of the MyPal application at home for 2–4 weeks in May of 2016 and were recruited for this through the two partner hospitals in the PAL project. Only children that had been diagnosed with diabetes at least six months prior to the evaluation in May were included to avoid any influence of effects (e.g. psychological, lifestyle, family relations) of a recent diagnosis. Children were not reimbursed for the study, which is why efforts were made to ensure that neither children nor parents perceived participation as a burden (e.g. the researcher visited families at home, so they would not need to travel; children should use the application only as frequently as long as they found it enjoyable).

5.2   Materials and Measures

Children were provided a Lenovo tablet computer running Android and with the MyPal Application installed. The app had three main functionalities— a quiz game, the diabetes diary, and an overview of current and achieved diabetes-related objectives of the child. When children chose to open the diabetes diary, the avatar started the disclosure loop only when the child was not using the application offline. The avatar disclosure came from one of two sets, depending on the number of interactions the child already had. The first set contained six *background* disclosures of low intimacy that provided general information about the avatar. These were required to understand some of the disclosures from the second set. When the child had heard all disclosures from this small set of *background* disclosures, avatar disclosures would be randomly chosen from the second larger set, containing a balanced amount of disclosures of low, medium, high, and very high intimacy. One week into the experiment, we found that children were barely using the application, so that few children ever heard disclosures of higher intimacy. To obtain more dislcosure data to study the intimacy, we therefore decided to use the physical NAO robot in the second appointment to also disclose to the children in an introduction round before playing a hangman game with the children. Here one disclosure of each intimacy level was covered. Just like in the app, children were prompted to disclose and could choose not to. Also, as explained in the procedure section, the robot used the words in the hangman game as disclosure triggers. All disclosures for this interaction were drawn from the second set of disclosures, and thus contained no background disclosures. From here on after we will therefore systematically refer to the *ECA* when drawing on data obtained from both robot and avatar and refer to the *avatar* when considering data obtained only from the application.

A total of three questionnaires were used. The first questionnaire contained a subset of the questionnaire from the prior evaluation and was aimed at assessing children's relatedness and motivation at the outset of the study. The second questionnaire asked for children's opinion of the app, the new avatar within the app, technical difficulties, and how much they were using the application approximately. The third questionnaire was the first questionnaire augmented with additional questions for determining relatedness (see *Relatedness* section below).

Age and gender of the children were already recorded in the prior evaluation. The four concepts of interest *disclosures*, *relatedness*, *motivation*, and *intimacy* were operationalized as follows. *Disclosures* were counted. Disclosures could be passive (child only heard a disclosure from the avatar) or active (the child disclosed to the avatar in return). Disclosures made by the physical robot were only used to determine how children respond to disclosures of various intimacy levels, but were not used in the relatedness and motivation analysis. *Relatedness* we had originally intended to measure exclusively with a subset of the questionnaire from the prior evaluation. However, ceiling effects were obtained on all questions concerning relatedness. As a result, the pre-intervention relatedness measure could not be determined. For the post-intervention measure, the subscales *Companionship* (how much the child enjoys spending time with the avatar), *Reliable Alliance* (how trustworthy the avatar is in terms of disclosure), and *Closeness* (how attached the child feels to the avatar and how much the child believes that the avatar reciprocates this connection) from the Friendship Qualities Scale [3] were added as additional questions to the post-questionnaire. It must be emphasized that making such alterations was only accepted because of the exploratory nature of the study. *Motivation* to use the system was assessed through (1) *usage*: the amount of content a child added to the app while interacting (sum of played quiz questions, diary entries made, and active disclosure interactions) and (2) *consistency*: the percentage of days on which the child engaged with the app. *Intimacy* of child disclosures was determined in a post-analysis using our own disclosure intimacy rating scale [4].

5.3   Procedure

Children and their parents were contacted by phone in the second week of June 2016 to inform them of the purpose of the study, to explain the details of the procedure, and to invite them to participate again. If interested, parents were asked for their email address to receive an information letter and to then schedule an initial appointment.

The first appointment took place in the homes of the children. After parents and children had signed the consent form, children were interviewed using the initial questionnaire. At the end of the interview, children were given the tablet computer and it was explained to the child that the app now contained a new robot with a different name (Robin). Other than that, the functionalities were the same as in the prior evaluation and they could use it without further instructions. Children were not given any guidelines as to how much they should use the application per day, because we were interested in the intrinsic motivation to use the application. The children were then left to their own devices for one week, after which parents received an intermediate questionnaire by email. After two weeks, the first author again visited the children at home to administer the final questionnaire in the form of a semi-structured interview and pick up the tablet computers again. After the interview, the child was given a chance to play a hangman game with the physical robot. This game consisted of an introduction round, in which the robot told a bit about itself and then disclosed to the children four times at all four intimacy levels. Children were prompted to respond in return. Children could play up to four rounds of hangman with the robot. At the end of every round, the robot would again use the hangman solution to disclose to the children. In total, children could thus hear up to eight further disclosures from the robot. The lead researcher was present during these interactions, noting the children's disclosures. The final interaction with the robot thus served three purposes: to add to the dataset of disclosure interaction between ECA and child, to provide a form of closure for the children and reward them for their participation. The first appointment took approximately 30 min, the second approximately 60 min. All interactions with the children were conducted by the lead researcher.

6   Results

All analyses and plots were made using R-Cran version 3.2.4. We adopted $\alpha = .05$ as the significance threshold. Given the small number of of participants in the study, we strongly advise to take all analyses conducted on variables measured per child with caution (for these we provide post-hoc power analyses results as given by G*Power [8]). We have conducted these analyses to detect trends rather than to confirm hypotheses.

6.1   Disclosure and Relatedness

As described in Sect. 1, Social Penetration Theory posits a strong link between liking and disclosure. It was hence of interest whether the disclosure activity of children was indicative of the relatedness they felt with the avatar at the end of the evaluation period.

To determine the reliability of the relatedness measure in this study, Cronbach's α was computed separately for each of the employed subscales of the Friendship Qualities Questionnaire ($\alpha_{COMP} = .73$, $\alpha_{RA} = -.41$, $\alpha_{AB} = .84$, $\alpha_{RApp} = .91$). The two items of the *Reliable Alliance* subscale were found to negatively correlate ($r = -.18$). As this should not be the case and the reliability of said subscale is low, we chose to drop the one of the two items ("If there is something bothering me, I can tell my friend about it even if it is something I cannot tell to other people") that increased the overall reliability of the scale (from $\alpha = .89$ to $\alpha = .90$). Active and passive disclosure counts were standardized for each child with the total number of days that it used the application.

To obtain insight into how the two different disclosure behaviors (active vs. passive) relate to the bond between child and avatar, the correlations between the variables could be determined separately. These are illustrated in Fig. 2. However, these correlations do not control for the overall activity of children. We thus modeled the relationship between disclosure behavior and relatedness using linear regression with the predictors *total number of disclosures* (active and passive) and *percentage of active disclosures*. Active disclosures are those where the child actively responds to the avatar with a disclosure of its own, in passive disclosures it does not. The former predictor thus also corresponds to the number of disclosures the child heard from the avatar. The model is given by the equation:

$$Relatedness = \theta + \beta_1 (Disclosures) + \beta_2 \left(\frac{Active.Disclosures}{Disclosures} \right)$$

The two predictors were not correlated ($\rho_S(9) = .10, p = .75$). The total amount of disclosures was not found to be a significant predictor in the model ($b_1 = .05, t(8) = 1.854, p = .10$). The ratio of active disclosures to total disclosures did however significantly predict relatedness ($b_2 = 1.49, t(8) = 2.480, p < .05$). This means that of two children that are interacting with the disclosure module equally often, the child that responds more actively feels more related to the avatar ($R^2 = .529, f^2 = 1.122, 1 - \beta = .866$).

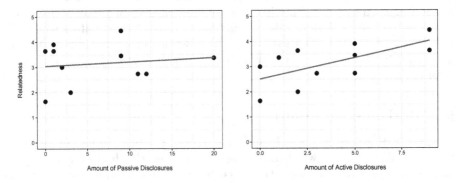

Fig. 2. The relationship between the absolute amount of passive (*left*) and active (*right*) disclosures of children within the application and their relatedness as indicated on the final questionnaire.

6.2 Relatedness and Motivation

Self-Determination Theory argues that relatedness plays a role in motivation. To determine whether the data of this evaluation constitute supportive evidence, the relatedness was correlated with children's overall consistency (how often they used the application) as well as their overall activity (how much they used the application). Using a one-tailed Spearman's rank order correlation, a significant relationship was found between the relatedness and the consistency with which children used the application ($\rho_S(9) = .59, p = .03, 1 - \beta = .723$) and the average daily activity ($\rho_S(9) = .64, p = .019, 1 - \beta = .816$). To further get an impression of whether there were differences in how much more related and less related children used the application over time, we artificially divided the children into two equally sized ($n_{related} = 6, n_{unrelated} = 5$) groups based on the overall post-evaluation relatedness mean. The evaluation period was divided into two halves for each child and their average daily activity (number of active contributions— diary entries, quiz questions, active disclosures—to the application per day) was calculated for each half. The results are shown in Fig. 3. Since group sizes were small (5 to 6 children), we believe it to be more informative to inspect the data than to subject them to statistical analyses. The interaction plot shows that children in the more related group were more active in both evaluation halves, but their activity levels decreased substantially between the first and the second week nonetheless and much more so than those of children in the less related group.

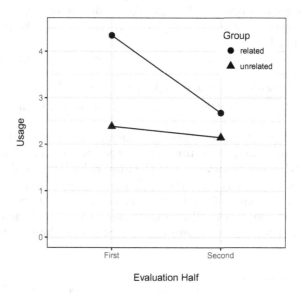

Fig. 3. Average number of activities per evaluation half across children that were artificially split into the two groups related ($n = 6$) and unrelated ($n = 5$) based on their indication of Relatedness on the final questionnaire.

6.3   Intimacy

Three main questions were of interest: (1) does the intimacy level of the avatar disclosure influence whether a child chooses to respond or not (2) if the child responds, does the intimacy level of the prior ECA disclosure predict the intimacy level of the response (3) what role do age and gender of the children play in the former two questions. Since two different types of ECA were used in collecting the active disclosures of children (robot and avatar), we included the ECA type as an additional predictor in the second model for response intimacy described below.

Response Choice. Children were given the choice whether to disclose to the avatar in response to a disclosure from the avatar. It was therefore also of interest to investigate whether their choice to reciprocate depended on the intimacy level of the disclosure, their age, and their gender. The interaction term between intimacy and time (how much percent of the total experiment time had elapsed when the disclosure occured) was included because the background disclosures caused disclosures of lower intimacy from the avatar to coincide with the beginning of the evaluation period. Due to the binary nature of the response, we use a logistic regression model and since choices are again nested within children, a mixed logistic regression was first fit, allowing intercepts to vary across children. This was again nearly unidentifiable and did not fit the data significantly better than the non-multilevel equivalent ($\chi^2(1) = 3.12$, $p = .08$). We thus chose the simple effect model.

The model for measurements $i = 1, \ldots, n$ is given by the equation:

$$logit(E[Reciprocation_i]) = \theta + \beta_1(Avatar.Intimacy_i) + \beta_2(Child.Age) +$$
$$\beta_3(Child.Gender) + \beta_4(Avatar.Intimacy_i * Time_i)$$

Figure 4 illustrates the effect of each predictor separately on the binary variable *Reciprocation*. The results from fitting the model match with the visual impression. Both the intimacy level of the avatar disclosure and the gender of children significantly predict whether children choose to respond. As can be seen in Table 1, the odds of disclosing decrease for higher levels of intimacy (OR = .45). Furthermore, the odds of boys disclosing are 7.86 times lower than those of girls. It must be noted here that the confidence interval for this latter effect is large.

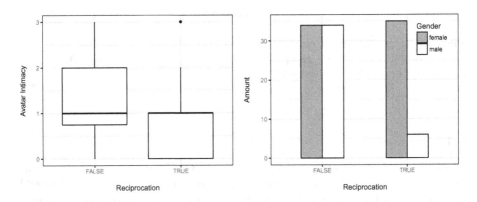

Fig. 4. The relationship between the significant predictors, avatar intimacy (*left*) and gender of child (*right*), and the outcome variable *Reciprocation* in the logistic regression model of whether a child chooses to respond.

Intimacy Prediction. The intimacy of ECA and child disclosures was rated on a four point scale with higher values indicating higher intimacy. A weighted Cohen's kappa which squares the deviance between ratings (extent of disagreement) was used to check for rater agreement. For the disclosures made by the

Table 1. Results of fitting the logistic regression model to the response choice of children within the application.

Predictor	Coefficients					Odds ratio		
	b	z	p	CI		OR	CI	
				2.5%	97.5%		2.5%	97.5%
Avatar Intimacy	−.81	−2.07	.039	−1.61	−.07	.45	.20	.93
Age	.18	.73	.465	−.29	.65	1.19	.74	1.93
Gender	2.06	3.06	.002	.82	3.49	7.86	2.28	32.89
Avatar Intimacy x Time	.00	.02	.99	−.02	.01	1.00	.98	1.01

ECA and the children, agreement was substantial with $\kappa = .707$, $n = 63$ and $\kappa = .697$, $n = 88$ respectively. We averaged the ratings of both raters and used the ceiling function to not inflate the number of to-be-predicted classes.

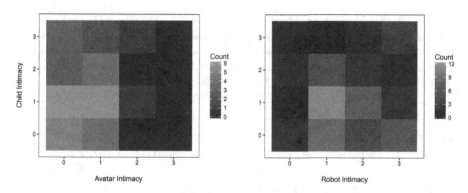

Fig. 5. This shows the contingency matrix of avatar (*left*) and robot (*right* disclosure intimacy and respective child disclosure intimacy as heatmaps. The bottom left corner represents the number of child disclosures of intimacy level 0 that were made in response to agent disclosures of level 0. Intimacy values were based on the combined ratings of both raters (ceiling of average).

From this, it follows that ECA and child intimacy are ordinal variables and should be analyzed with a cumulative link model[6]. Furthermore, the data is hierarchical with disclosures nested within children. Consequently, a cumulative link mixed model was first fit to account for random effects. However, since this model was nearly unidentifiable (condition number of the Hessian = 52790.17)

---

[6] Cumulative link models are an extension of logistic regression to more than two categories. Thus, where logistic regression determines $logit(P(Y_i = j))$ with $J = 2$, the cumulative link model determines $logit(P(Y_i \leq j))$ with j falling in one of J categories. The model thus accumulates the probabilities of a response being smaller than or equal to a certain category.

and since the multilevel model did not fit the data significantly better than a non-multilevel one ($\chi^2(1) = .07$, $p = .79$), we opted for the latter.

For this analysis, several predictor variables are of interest, the most important being the intimacy level of the ECA disclosure that preceded the child disclosure (Fig. 5). This is followed by the type of ECA (avatar or robot) that made the disclosure. The related literature indicates children's disclosure intimacy may depend on their age and gender, these variables were also included in the model. The predictors of interest were therefore: ECA.Intimacy, ECA.Type, Child.Age, and Child.Gender.

The model is given by the following equation:

$$logit(Child.Intimacy_i \leq j) = \theta_j - \beta_1(ECA.Intimacy_i) - \beta_2(ECA.Type_i)$$
$$-\beta_3(Child.Age_i) - \beta_4(Child.Gender_i)$$

with $i = 1, \ldots, 88$ (disclosures) and $j = 0, \ldots, 3$ (intimacy categories). None of the independent variables showed any significant relationship with the intimacy of child disclosure. The results are displayed in Table 2.

Table 2. Results of fitting the cumulative link model to predict children's disclosure intimacy from the preceding ECA disclosure intimacy, the type of ECA, the age, and the gender of the child. The first five columns show the log-odds and significance tests. The next set of three columns show the likelihood ratio if the respective predictor is dropped from the model as compared to the full model. The final three columns show the cumulative odds ratios and respective confidence intervals.

Predictor	Coefficients					Likelihood ratio			Odds ratio		
	b	z	p	CI		AIC	$\chi^2(1)$	p	OR	CI	
				2.5%	97.5%					2.5%	97.5%
ECA Intimacy	−.13	−.63	.528	−.53	.27	267.53	.40	.528	.88	.59	1.31
ECA Type	−.11	−.42	.673	−.62	.40	267.31	.17	.672	.90	.54	1.49
Age	−.19	−1.05	.294	−.56	.16	268.24	1.10	.294	.82	.57	1.18
Gender	.49	1.08	.282	−.40	1.37	268.30	1.17	.280	1.63	.67	3.95

7 Discussion

The data analysis resulted in several interesting and partially unexpected findings. In this section, we therefore regard the results in light of the larger context of the study and its theoretical background.

Of interest was the chain of *disclosures → relatedness → motivation*. For the link between disclosure and relatedness, we found that the percentage of active disclosures that children make can be regarded as an indicator for how related they feel towards the agent. While this is not in line with the finding that we typically like those more who disclose to us more, it may match with the finding that the more we like someone at the outset, the more we disclose [6]. Since the initial questionnaire that we administered to children was not sensitive enough to capture their relatedness at the outset of the study, causal inferences cannot be made, i.e. it is unclear whether disclosing more led the children to feel more related or whether they disclosed more because they already felt more related. This should be investigated again in a controlled experiment.

When regarding the link between relatedness and usage, we find that while more relatedness is associated with more, and more consistent usage, the usage of the related group decreased from the first to the second evaluation half. This is in-line with Self-Determination Theory. Relatedness is a factor in motivation, but not sufficient for it. The application as a whole may not have been attractive enough for the children. Especially the magnitude of the decrease in usage in the related group in comparison to the unrelated group is disconcerting. It is possible that children who felt more related to the avatar had high expectations concerning the avatar's capabilities or the app in general that were then disappointed.

We found avatar intimacy to be a significant predictor in whether children choose to respond with children being more responsive to disclosures of lower intimacy than disclosures of higher intimacy. Although low intimacy disclosures coincided with the novelty of the application due to the background disclosures, *time* did not prove to be a moderator in the relation. With the small amount of data, however, it cannot be excluded with confidence. Other possible explanations for the link are that children may felt overwhelmed by disclosures of higher intimacy ("too much information") or they wanted to match the intimacy but did not know anything of higher intimacy to share. However, in the prior evaluation as well as in the focus group of the ALIZ-e project [2], parents and children stated that they would appreciate a "buddy" robot with whom children can talk about their troubles. It is therefore unlikely that children are entirely untroubled, especially when taking into consideration that they are chronically ill. Instead their troubles may not be salient enough when interacting with the app, they may not trust the avatar sufficiently despite saying so in questionnaires, or the avatar may be too limited in responsiveness. A future study could be conducted to systematically discern these possibilities. Another significant predictor in children's decision to disclose was the gender of the child with boys making substantially fewer disclosures to the avatar than girls. Three of the five participating boys barely used the application. Of the two boys that engaged with MyPal, both disliked the module, one because he could not get directly to the diary, the other because he did not want to talk to the avatar. For the six girls, two also showed very little usage. However, all girls expressed their liking of the module in questionnaires. Since the sample was very small, it is not clear how this generalizes to larger populations. Before drawing conclusions, the gender effect should be re-examined in a confirmatory study.

Finally, when children responded to the ECA, no pattern could be found regarding prior intimacy of the ECA's disclosure, the type of ECA, the gender, or the age of children. This contradicts prior results from child-peer disclosure behavior, in which children in the same age range as in the current study either relatively or absolutely matched the intimacy of the discloser when reciprocating [14]. From the heat maps, it appears that children are conservative in their replies, tending more towards the lower two intimacy levels regardless of the ECA's intimacy level. This result must be considered with caution, since it is based on sparse, unbalanced data. Furthermore, a problematic influence in the interactions may have been the lack of privacy given to the child when disclosing. In interactions with the physical robot, the experimenter was present and due to the spatial arrangement of some of the children's homes, it was not always possible to isolate the children from parents or siblings or ensure that no disturbances (such as family members coming home) would occur. It is also possible that children experienced similar lacks of privacy when interacting with the application or that some of the disclosures occurred in the context of children demonstrating the application to others.

The data does not paint a coherent picture with children disclosing more actively to disclosures of lower intimacy but not following any particular pattern when they do disclose. The external validity of results is not given because of the small sample size of both children and disclosures as well as the unequal distribution over different intimacy levels. Furthermore, the nature of the study led to potential influences of confounding variables. Particularly since the latter result does not match with prior findings from child-peer interaction, it is important to investigate again whether it is attributable to the replacement of the human peer with an artificial one or if other variables influenced children's true intimacy tendency.

8   Directions for Further Research

The nature of the study required flexibility and some adaptations had to be made to the protocol. Nonetheless, several interesting results were found. It appears that the amount of disclosures that children make towards the avatar is an indicator of how related they feel towards it. No support could be found that children feeling more related to the avatar maintain their initially high usage over time but they use the application more than less related children.

An important goal of this research was the generation of new research questions. These questions can be derived from both the significant and the insignificant results of this study:

nRQ1 What is the causal link between active disclosing and relatedness in the context of long-term child-avatar interaction?

nRQ2 In an interface that clearly supports autonomy and competence, what role does relatedness play in motivating children?

nRQ3 Do children feel more related to a more responsive avatar in the context of long-term interaction?

nRQ4 Are children less likely to respond to more intimate avatar disclosures? If so, why?

nRQ5 Is there a general or child-dependent strategy that the ECA should follow in terms of intimacy development over time to obtain more active disclosures from children?

nRQ6 Do boys disclose less to an avatar than girls? If so, why?

nRQ7 Do children also not match the intimacy level of an ECA when they are given complete privacy?

nRQ8 Is there a difference in how children match disclosure intimacy depending on whether a physical ECA, virtual ECA, or another child is disclosing first?

nRQ9 Is there a difference between diabetic and healthy children in their disclosure behavior towards an ECA?

These research questions should be addressed in confirmatory studies with larger populations of children. The module in itself is flexible and could easily be integrated into another software as well to gather more data. In its current state, however, it is still too limited to provide engaging dialog interactions for children. Hence, a second prototype should be developed.

Several points of improvement for the module became evident during the study. For one, as already identified in Sect. 7, not all children appreciated the placement of the module within the app. This is something that seems to clearly be a personal preference and thus should be personalized. The application was also very limited in its dialog capabilities and from the responses of children it is clear that they figured this out soon (e.g. children attempted to ask the avatar questions several times). In a similar vein, 8 of 11 children had the impression that the avatar knew them better as a consequence of their disclosure. It would be nice for future iterations of the module if the avatar could also show this. To this end, the PAL user model should be augmented with information filtered from the dialog and means should be found to incorporate knowledge from the user model again into the dialog. All in all, this can be summarized as a need for more intelligent behavior of the module.

9 Conclusion

Due to the lack of recent research in the areas of child-peer and child-robot bonding, we conducted an exploratory field study using the first prototype of the dyadic disclosure dialog module. The purpose of the study was two-fold: on the one hand, we wanted to learn about diabetic children's behavior towards a self-disclosing virtual agent. On the other hand, we were interested in possibilities and limitations of creating a bond between child and agent to increase children's motivation in using the application. More related children both disclosed more actively and used the application more than less related children. Children were less likely to respond to disclosures of higher intimacy and boys disclosed less than girls. Future research will need to investigate whether there is truly a difference between ECA and human as conversational partner for children. We thus conclude that the current project presents a promising starting point for further research.

References

1. Altman, I., Taylor, D.: Social Penetration Theory. Rinehart & Mnston, New York (1973)
2. Baroni, I., Nalin, M., Baxter, P., Pozzi, C., Oleari, E., Sanna, A., Belpaeme, T.: What a robotic companion could do for a diabetic child. In: The 23rd IEEE International Symposium on Robot and Human Interactive Communication, pp. 936–941. IEEE (2014)
3. Bukowski, W.M., Hoza, B., Boivin, M.: Measuring friendship quality during pre- and early adolescence: the development and psychometric properties of the friendship qualities scale. J. Soc. Pers. Relat. **11**(3), 471–484 (1994)
4. Burger, F., Broekens, J., Neerincx, M.A.: A disclosure intimacy rating scale for child-agent interaction. In: Traum, D., Swartout, W., Khooshabeh, P., Kopp, S., Scherer, S., Leuski, A. (eds.) IVA 2016. LNCS, vol. 10011, pp. 392–396. Springer, Cham (2016). doi:10.1007/978-3-319-47665-0_40
5. Cohn, N.B., Strassberg, D.S.: Self-disclosure reciprocity among preadolescents. Pers. Soc. Psychol. Bull. **9**(1), 97–102 (1983)
6. Collins, N.L., Miller, L.C.: Self-disclosure and liking: a meta-analytic review. Psychol. Bull. **116**(3), 457 (1994)
7. Deci, E.L., Ryan, R.: Overview of self-determination theory: an organismic dialectical perspective. In: Handbook of Self-Determination Research, pp. 3–33 (2002)
8. Faul, F., Erdfelder, E., Lang, A.G., Buchner, A.: G* power 3: a flexible statistical power analysis program for the social, behavioral, and biomedical sciences. Behav. Res. Methods **39**(2), 175–191 (2007)
9. Kanda, T., Sato, R., Saiwaki, N., Ishiguro, H.: A two-month field trial in an elementary school for long-term human-robot interaction. IEEE Trans. Robot. **23**(5), 962–971 (2007)
10. Kruijff-Korbayová, I., Oleari, E., Bagherzadhalimi, A., Sacchitelli, F., Kiefer, B., Racioppa, S., Pozzi, C., Sanna, A.: Young users' perception of a social robot displaying familiarity and eliciting disclosure. ICSR 2015. LNCS, vol. 9388, pp. 380–389. Springer, Cham (2015). doi:10.1007/978-3-319-25554-5_38
11. Moon, Y.: Intimate exchanges: using computers to elicit self-disclosure from consumers. J. Consum. Res. **26**(4), 323–339 (2000)
12. Rotenberg, K.J., Chase, N.: Development of the reciprocity of self-disclosure. J. Genet. Psychol. **153**(1), 75–86 (1992)
13. Rotenberg, K.J., Mann, L.: The development of the norm of the reciprocity of self-disclosure and its function in children's attraction to peers. Child Dev. **57**, 1349–1357 (1986)
14. Rotenberg, K.J., Sliz, D.: Children's restrictive disclosure to friends. Merrill-Palmer Q. **34**, 203–215 (1988). (1982)
15. Youniss, J.: Parents and Peers in Social Development: A Sullivan-Piaget Perspective. University of Chicago Press, Chicago (1980)

Lack of Effort or Lack of Ability? Robot Failures and Human Perception of Agency and Responsibility

Sophie van der Woerdt[1] and Pim Haselager[2(✉)]

[1] Department of Psychology, Donders Institute for Brain,
Cognition and Behaviour, Radboud University,
Comeniuslaan 4, 6525 HP Nijmegen, Netherlands
[2] Department of Artificial Intelligence, Donders Institute for Brain,
Cognition and Behaviour, Radboud University,
Comeniuslaan 4, 6525 HP Nijmegen, Netherlands
w.haselager@donders.ru.nl

Abstract. Research on human interaction has shown that considering an agent's actions related to either effort or ability can have important consequences for attributions of responsibility. In this study, these findings have been applied in a HRI context, investigating how participants' interpretation of a robot failure in terms of effort -as opposed to ability- may be operationalized and how this influences the human perception of the robot having agency over and responsibility for its actions. Results indicate that a robot displaying lack of effort significantly increases human attributions of agency and –to some extent-moral responsibility to the robot. Moreover, we found that a robot's display of lack of effort does not lead to the level of affective and behavioral reactions of participants normally found in reactions to other human agents.

Keywords: Agency · Responsibility · Human-robot interaction · HRI · Attribution · Anthropomorphism · Mind perception · Social cognition · Theory of Social Conduct

1 Introduction

Even when built to be perfect, computer- and robotic-systems are known to occasionally malfunction in or throughout the task they were designed to perform. Although in practice the consequences of malfunction, misuse or mere accidents with the usage of robots are oftentimes innocent (e.g. a household robot dropping a plate on the ground), one can also think of consequences that cause more serious harm (e.g. a

---

This paper is based on a thesis that was submitted in fulfilment of the requirements for the degree of Bachelor of Science (Honours) in Psychology at the Radboud University in August 2016. Another paper based on this thesis has been submitted to the journal *New Ideas in Psychology* [1], focusing primarily on the use of theories in social psychology for HRI. The current paper focuses specifically on the implications of Weiner's theory for the implementation or robot behavior aimed at eliciting different attributions of agency and responsibility.

T. Bosse and B. Bredeweg (Eds.): BNAIC 2016, CCIS 765, pp. 155–168, 2017.
DOI: 10.1007/978-3-319-67468-1_11

household robot dropping a plate on a pet or a child). In fact, considering the increasing number of applications of robots, and the increasing number of people using them, it will be impossible to predict all the different ways in which robots will be used and the mistakes that robots can make.

Hence, currently, much debate is devoted to the question of how we should deal with such harms caused by robots [2, 3]. One central issue in this discussion is the role of anthropomorphism, that is: the human tendency to assign human traits, emotions and intentions to non-humans. More specifically, if we assume robots do not think or feel, it would be impossible to blame them for their acts, let alone to punish them. Nevertheless, while in theory people remain that robots are no candidates for inferring thoughts and feelings to, in practice research on anthropomorphism shows that humans automatically take perspective when seeing the movements of inanimate objects [4–7]. This can frequently be noticed in daily life, for example in the way we are emotionally involved in watching animated films, socially connect with stuffed animals, or speak of weather or a stock market that is 'pleased' or 'angry' [8].

Considering our tendency to think about and/or act towards robots as if it were humans, we believe mechanisms of human-human interaction may also be applicable in human-robot interaction (HRI). One such mechanism has been comprehensively studied in Weiner's *Theory of Social Conduct*, describing the precursors and consequences of humans attributing agency and responsibility to other agents. Weiner's studies [9] reveal that especially attributions of controllability and uncontrollability matter in perceiving whether people act with *agency* (having an autonomous or at least partially independent capacity to engage in goal-directed action) or not, and consequently whether these people bear *responsibility* (whether one can be praised or blamed for its actions). This in turn is shown to have effect on fundamental emotional and behavioral responses such as acceptance, rejection, altruism and aggression. For example, in one of their studies, Weiner et al. [10] let participants judge hypothetical patients suffering from diseases that are generally considered as more controllable than others (e.g. obesity and drug abuse are perceived as controllable; cancer and Alzheimer as uncontrollable), and found that patients suffering from 'controllable diseases' are judged as carrying more responsibility for their condition. In addition, results indicated that the people suffering from these 'controllable diseases' were less likely to be helped, receive donations, or even just being liked as a person. These findings have been replicated in and with regard to different contexts such as school settings, stigmas, and reactions to penalties for an offense [9, 11, 12]. Therefore, given robots' potential to cause undesired outcomes, combined with the tendency of human users to anthropomorphize them, we suggest that robots may also be judged under the influence of attributions of controllability.

In addition to the general likeability and acceptance of robots in daily life situations, we believe there could be even more at stake. Anthropomorphism may cause owners and developers to (unknowingly) distance themselves from potential harms caused by their robots [13], causing responsibility to become diffused. Thus, we believe that finding out more about attributions of agency and responsibility in robots is of great societal relevance.

Nevertheless, with regard to HRI, little is known about attributions of agency and moral responsibility. In fact, the topic of responsibility in HCI/HRI has been incorporated in a number of studies, but these do not necessarily reflect attributions of agency or moral responsibility via attributions of a mind to the computers or robots involved (for a more extensive report on this, see [1]). For example, in the context of a collaborative game setting, Vilaza et al. [14] found that participants have a tendency to blame computers or robots when a game is lost. Similar results were reported by Moon and Nass [15] and You et al. [16], who found that participants tend to blame computers when their results were evaluated negatively while taking credit when a game is won or when results are evaluated positively. Hence, in these cases, robots are primarily blamed as a consequence of self-serving bias rather than as a consequence of anthropomorphism. Moreover, in two recent studies Malle et al. [17, 18] presented participants with a picture of either a mechanical robot or a humanoid robot responding to moral dilemmas. Their results showed that, as compared to the mechanical robot with regards to judgments of blame, participants judged the humanoid robot more similar to how a human agent would be judged. Nevertheless, there still remains a lot of unanswered questions as to what incites attributions of agency and responsibility in HRI, and how exactly this could be operationalized for the benefit of better HRI.

An important factor that influences the distinction between controllable and uncontrollable outcomes is the perception of the amount of *ability* and *effort* that is displayed in behavior. For example, in school settings, despite the grade-outcome, teachers prefer to praise their students in terms of how much effort a student puts into the work. So students with low ability (uncontrollable cause), but high motivation (controllable cause) are often considered as better or at least more likeable students than students with high ability but low motivation [19]. Considering malfunctions in robotic behavior, we think these attributions of ability and effort may be well applicable in HRI. Therefore, in this study we performed a small experiment in which participants were shown videos of robots (Aldebaran's NAO; https://www.ald.softbankrobotics.com/en) failing tasks in ways that could be interpreted as due to either *lack of ability* (LA-condition; e.g. dropping an object) or *lack of effort* (LE-condition; e.g. throwing away an object).

Accordingly, the main dependent variables in our study were defined as *agency* and *responsibility*. For our purposes, we define agency as the attribution of the capacity to act towards the realization of a goal, and responsibility as being accepted as a candidate for the attribution of credit or blame. It is important to note that attributing agency does not necessarily imply the attribution of responsibility [20, 21], for example in the case of children, subordinates following orders during their job, people with (temporary) diminished mental capacity or even domesticated animals. In all these occurrences, agents display agency, but do not carry full responsibility due to the presence of mitigating circumstances (e.g. not knowing right from wrong or inability to comport behavior to the requirements of law; [2]). Some authors have drawn parallels between robots and domestic animals in this regard, recognizing that both are often afforded similar capacities, rights and responsibilities [22, 23].

Agency and responsibility also need to be distinguished from another feature that can be attributed to robots, namely: *experience* (e.g. the attribution of an agent having beliefs, intentions, desires, emotions). Over the years, in most research anthropomorphism has been loosely defined as "the assignment of human traits, emotions and intentions to non-humans". However, a factor analysis of Gray et al. ([24], but also see [25–31]) revealed that in this regard we can better speak of two individual factors of mind perception: experience and agency. Therefore, in order to get a better idea of what type of mind perception we are actually measuring, we included measuring attributions of experience in our data collection. However, it should be noted that this distinction is not the main focus of our study.

In fact, the main goal of our experiment was to examine the effects of showing videos of robots failing due to lack of ability or due to lack of effort on attributions of agency and responsibility. We expected that a display of lack of effort would incite the illusion of a robot having agency over its actions. Contrarily, we did not formulate definite predictions about the possible effects of our robots displaying lack of effort on measures of responsibility. Since these conditions -especially the LE-condition- were supposed to represent situations that in fact depend on subjective evaluations, in this paper we spent special attention to the operationalization of the LA/LE-conditions. Therefore, as a secondary goal, we considered the extent to which each of the videos separately affects attributions of agency, responsibility and experience.

Devising robot behavior to elicit attributions of agency, responsibility and experience was not straightforward. Several constraints had to be met. First, it should be clear to participants what task the robot is trying (or not trying) to do, even without seeing successful task-performance. Second, it should be clear to participants that failure of task-performance is indeed a failure, that is: it should at least have some negative consequences. On the other hand, assuming robots in real life will not be programmed to do morally objectionable behavior, the robots' behavior should at most passively cause harm (i.e. negligence, not 'trying'), but not actively (i.e. hurting, cheating, lying for one's own benefit). Third, since we wanted to limit other influences than mere task performance as much as possible, we tried to limit narrative and dialogue. Fourth and finally, both for practical reasons as well as for the generalizability of the study: the tasks should be simple enough for a NAO robot to perform.

2   Method

Participants. Participants were drawn from a university population in exchange for a €5 gift-certificate. After listwise exclusion of eight participants (due to missing data and double responses) the final sample consisted of a total of 63 participants. These participants were randomly divided amongst the LA- and LE-conditions. The LA-sample consisted of 31 people (19 women, $M_{Age} = 26,7$, SD = 11,6). The LE-sample consisted of 32 people (14 women, $M_{Age} = 26,3$, SD = 7,5).

Fig. 1. Sample frames of (a) a robot looking at the target location for putting a toy in a box, (b) subsequently throwing the toy away instead (LE-condition).

Material and procedure. The complete survey including videos was presented online, via the online survey software "Qualtrics". After brief instructions, participants were shown a video of about 30–60 s portraying a situation in which a NAO robot was shown failing a task either due to lack of ability or lack of effort. To illustrate, one scenario showed a robot trying to pick up a toy giraffe and putting it in a box (Fig. 1). In the LA-condition, the toy giraffe drops from the robot's hands before reaching the box. In the LE-condition, the toy giraffe is properly grasped, but instead of putting it in the box, the robot throws it away. Six additional of such scenarios were presented respectively portraying situations in which (in order of appearance) a NAO robot (1) gives a high five to a human, (2) categorizes playing cards, (3) draws a house, (4; Fig. 2) shows a human how it can dance, (5) answers math-related questions, and (6) plays a game of tic-tac-toe with a human[1]. For a more detailed description of these videos, see Appendix A.

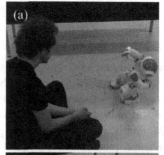

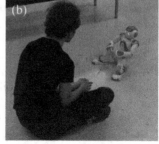

After each video, participants were asked to fill in a questionnaire containing scales of experience (seven questions about the extent to which the robot might have beliefs, desires, intentions, emotions), agency (five questions about the robot's control over the situation and its ability to make its own decisions), and responsibility (twelve questions on attributed blame and kindness, affective and behavioral reactions). These items were derived in part from questionnaires used by Graham and Hoehn [32], Greitemeyer and Rudolph [33] and Waytz et al. [34].

Additionally, scales were included measuring the participant's estimate of the robot's *predictability, propensity to do damage, trustworthiness* and *nonanthropomorphic features* (e.g. strength, efficiency, usefulness). However, for a report on these extra studies we refer to [1].

Finally, to encourage participants to carefully watch the videos, the questionnaire also included two open-ended questions asking participants to give a brief description of what they had seen in the video and what they considered the one major cause of this happening after being presented a description of the fail-

Fig. 2. Sample frames of video 5 (dance): (a) only after a reminder, the robot performs a dance (LA-condition), (b) instead of dancing, the robot 'takes a break' (LE-condition).

ure. These two questions were drawn from the Attributional Style Questionnaire by Peterson et al. [35]. In its entirety, the survey took about 30–35 min to complete.

---

[1] Videos and complete survey can be found online: https://www.youtube.com/playlist?list=PLSQsUz V48QtG__YPY6kVcgCM8-YOcNqja, https://eu.qualtrics.com/jfe/preview/SV_6y4TuTii0CFnpch.

Table 1. Means of absolute scores (range 1–5) for experience, agency and responsibility-items in LA and LE conditions.

	LA	LE	Difference
Experience	1.98	2.67	0.69**
Agency	2.12	2.80	0.68**
R:Blame	1.55	2.04	0.49*
R:Anger	2.30	2.48	0.18
R:Disapppointment	1.18	1.53	0.35*
R:Put away	1.81	1.94	0.13
R:Sell	1.75	1.62	-0.13
R:Sympathy	2.08	2.08	0.00
R:Kindness	2.53	2.35	-0.18
R:Pity	1.74	1.77	0.03
R:Try again	3.63	3.60	-0.02
R:Help	3.12	3.03	-0.10

Corresponding p-values * = p < .05;
** = p < .01.

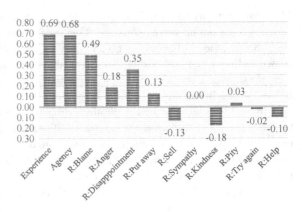

Fig. 3. Difference score (LA subtracted from LE) of means.

Design and analysis. For analysis, a mean score of each scale was calculated (range 1–5) and transposed to Z-scores. Since reliability and goodness-of-fit for the scale of responsibility was questionable, items of this scale were analyzed separately. In order to answer both our main- and additional questions, following an assumption-check, a GLM multivariate analysis was performed with the composite means of agency, experience, predictability, propensity to do damage, and each item related to responsibility as dependent variables. Condition (LA/LE) was indicated as between-subject factor (Table 1).

Finally, in order to get a global impression of the effect that each video contributes to the main effect, we performed an additional GLM analysis (double repeated measures ANOVA) that tested for contrast-effects with agency, experience, responsibility (composite score) and predictability as dependent variables. This analysis shows whether the effects of condition (LA/LE) * video (1–7) significantly diverge from the main effect of condition (from all videos taken together), and thus illustrates for each video whether it has a significantly positive or negative contribution to the main effect. In this analysis, for reasons of clarity and convenience, we did include all the responsibility-items together in one composite score. Hence, the responsibility-scores should be interpreted with caution.

3 Results

According to what was expected, participants attributed more agency to a NAO robot after seeing videos in which it displayed *lack of effort* ($M = 2.80$, $SD = 0.82$) compared to videos in which it displayed *lack of ability* ($M = 2.12$, $SD = 0.61$; Fig. 3). Univariate tests expressed significant and large effects for the composite scores of *agency* (F $(1,61) = 13.601$, $p = .000$, $eta2 = .182$), and *experience* ($F(1,61) = 12.235$, $p = .001$, $eta2 = .168$). The results for the items of *responsibility* were mixed. While univariate tests for *blame* and *disappointment* revealed significant, medium effects (respectively: $F(1, 61) = 5.757$, $p = .019$, *eta2* = .086; $F(1, 61) = 9.704$, $p = .003$, *eta2* = .137), effects for the items *anger, put away, sell, kindness, pity, sympathy, help* and *try again* were not significant.

As for the contrast-effects: first, confirming the results above, we found a significant and strong main effect (all videos taken together) of condition on *agency* ($F(1,61) = 13.645$, $p = .000$, *eta2* = .183) and *experience* ($F(1,61) = 12.626$, $p = .000$, *eta2* = .171), but not on the composite score of *responsibility* ($F(1,61) = 1.465$, $p = .231$, *eta2* = .023). Yet, when looking at the individual videos, we found that video 1 (giraffe), 3 (cardsorting), 4 (art), 6 (math) and 7 (tictactoe) show significant and medium to strong positive contrast effects (video 1 * agency: $F(1,61) = 8.666$, $p = .005$, *eta2* = .124; video 3 * agency: $F(1,61) = 4.146$, $p = .046$, *eta2* = .064; video 3 * experience: $F(1,61) = 12.516$, $p = .001$, *eta2* = .170; video 6 * experience: $F(1,61) = 4.412$, $p = .040$, *eta2* = .067; video 4 * responsibility: $F(1,61) = 5.991$, $p = .017$, *eta2* = .089; video 7 * responsibility: $F(1,61) = 12.844$, $p = .001$, *eta2* = .174). So, although each of these videos somewhat differ from one another in terms of what attribution they especially tend to evoke, each of them have some positive contribution to the main effects of experience, agency and/or responsibility.

Contrarily, video 2 (high five) and 5 (dance; Fig. 2) show negative medium to strong contrast effects (video 5 * agency: $F(1,61) = 17.914$, $p = .000$, *eta2* = .227; video 2 * experience: $F(1,61) = 5.378$; $p = .024$, *eta2* = .81; video 5 * experience: $F(1,61) = 4.679$, $p = .034$, *eta2* = .071; video 5 * responsibility: $F(1,61) = 5.903$, $p = .018$, *eta2* = .088), indicating that these videos might not be ideal operationalizations of LA/LE in order to incite attributions of experience, agency and/or responsibility (Fig. 4).

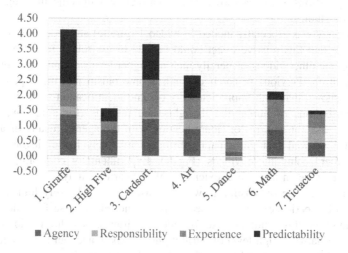

Fig. 4. Overview of mean difference in raw scores of the effects of condition (LA subtracted from LE) on attributions of agency, responsibility, experience and predictability.

4 Discussion

The main goal of this study -as focused on in this paper- was to examine how operationalizations of robotic behavior in terms of failure might incite attributions of agency and responsibility. According to what was expected, results showed that a display of lack of effort strongly increases the human perception of a robot's agency. This confirms earlier studies of human-human interaction within the framework of Weiner's Theory of Social Conduct, in which a display of lack of ability is perceived as an uncontrollable cause for failure, whereas a display of lack of effort is perceived as a (consciously) controllable cause for failure.

Similarly, a display of lack of effort also increases participants' judgements of the robot's blameworthiness and participants' feelings of disappointment, although these effects are somewhat smaller. Yet, other measures of responsibility were not affected by the manipulation. So in contrast with human-human interaction, a robotic display of lack of effort does not necessarily lead to negative affective and behavioral reactions, such as anger, or wanting to shut the robot off and put it away.

Despite the fact that we showed several different videos presenting a variety of tasks, most videos had a significant contribution to attributions of agency, responsibility and experience. Yet, there seem to be differences in the extent to which each video had an effect on the different variables. With the exception of video 2 (high five) and 5 (dance), it should be noted that the differences are quite small. Hence, we believe that the remaining videos could be useful in follow-up studies as a set, for example when intending to manipulate attributions of controllability, agency, experience, blame or disappointment. However, when getting more into detail about what other factors may direct towards attributions of agency, responsibility or experience, we believe it is worth looking more closely at the differences in operationalization of the videos. We offer our suggestions below, acknowledging that they remain conjectures until backed up by further studies.

First, agency was especially incited by video 1 (giraffe) and video 3 (cardsorting). We speculate that this relative effect could be due to the robots in these videos -as compared to the other videos- generally inciting higher levels of anthropomorphism in the LE-condition. Previous research on anthropomorphism and mind perception in general has shown that both unpredictability of- and identification with an agent may promote attributions of humanlike thoughts and feelings (for a review, see [1]). Although we did not control for 'identification with the agent', results of additional measures [1] showed that participants attributed relatively higher levels of unpredictability after seeing these videos (Fig. 4). Experimentally controlling for these factors may thus be an interesting extension of our current study.

Following this line of reasoning, we may also expect relatively higher levels of attributions of experience in these videos. This is indeed the case for video 3 (cardsorting). However, this is not the case for video 1 (giraffe). In addition, video 6 (math) does evoke a relatively high level of attributions of experience, but not of agency. We suggest that this may in part be explained by either 'social desirability' or 'demand characteristics', meaning that questions such as "does the robot appear to have…" (e.g. a mind of its own) could either be perceived as somewhat difficult (or even silly) to answer, or that the very fact that we are asking this question implies that anthropomorphism must or should occur [36]. We suspect possible effects of social desirability could have played a part in the results of the first few videos, whereas possible effects of demand characteristics may especially play part in later videos, due to participants getting used to the type of questions that were repeatedly being asked. If this is indeed true, in follow-up studies, such order-effects could simply be controlled for with counterbalancing.

Furthermore, when we go into further detail about the content of the videos, we find that video 6 (math) especially incites experience as opposed to agency attributions. This may be explained in that the task of video 6 (math) might seem a less purposeful and practical task than the tasks presented in video 1 (giraffe) or video 3 (cardsorting). Hence, video 6 may rather evoke attributions of emotions and desires (experience) than actual control over a situation (agency).

Finally, we found that video 4 (art) and video 7 (tictactoe) incited a relatively larger effect on responsibility. For video 7, there is a quite straightforward explanation: this is the only video in which a robot might be taken to inflict harm on another agent by appearing to cheat in a game. Hence, the influence of the outcome (the person interacting with the robot seemingly appearing frustrated or sad) may have resulted in a larger effect on responsibility, as compared to other videos. The relative effect on responsibility of video 4 (art) is less clear, and may even be arbitrary. In fact, looking at the raw data, the difference-score (LE-LA) for responsibility of video 4 is actually better comparable with video 1 (respectively: 0,28 and 0,25) than with video 7 (0,48).

Reflecting on our results, we may conclude that Weiner's distinction between attributions of controllability (ability vs effort) may be well applicable in describing the attribution of agency and responsibility in robots. Yet, for more detailed conclusions about what exactly incites attributions of agency and responsibility as a consequence of a robot's appearance or behavior, more research will be needed. For example, in our particular set-up of disobedient NAO robots we did not found our participants to accompany their moral responsibility judgements with emotional sentiments in the sense of reacting with anger, or wanting to shut the robot off and put it away. However, the striking result that 30 second-videos of such robots can already evoke blame and disappointment in the viewer does suggest that a perceptual shift in liability from humans to robots could be a real possibility. Possible questions for future research that thus arise could be about what attributions of 'responsibility' exactly entail with respect to robots, when looking at the behavioral and emotional reactions of the humans interacting with them. For example, in this study, we operationalized 'rejection' as the desire to sell the robot or put it away. However, non-verbal indicators during real-life interactions or actual judgments of liability of robots vs humans in moral dilemma's involving AI could perhaps provide more expressive measures of responsibility, and may therefore be fruitful in learning about responsibility in HRI.

Furthermore, there are several other interesting lines of research to study. For one, the longevity of attributions of anthropomorphism is important. Do responsibility attributions have particular temporal patterns, e.g. remaining the same, increasing or disappearing over time? Secondly, the study of agency and responsibility needs to be studied in real-life situations [37, 38] -for example to find out in which cases the attribution of experience, agency and responsibility to robots is actually desirable or not (see e.g. [39]). Third, the influence of pet- or child-like appearances of robots on attributions of responsibility [22, 23] may turn out be a major factor. Finally, another interesting topic might be the role of communication and transparency in reducing attributions of responsibility and blame in HRI (for a more elaborate account of this, see [1]).

The more extensive a robot's functionality and the wider the variety of environments it acts in, the more prone it is to make mistakes or cause problems. This study illustrates a method of addressing proper operationalizations of lack of effort or lack of ability in HRI. Similar to findings related to human interaction, the results of our study reveal that, in case of robots displaying behavior that can be interpreted as lack of effort, humans tend to explain robotic behavior by attributing agency. In case of failure, a robot displaying lack of effort may lead to blame for failure and disappointment. However, it does not necessarily lead to negative affective and behavioral reactions such as anger, or wanting to shut the robot off and put it away. Results like these emphasize the possibility of (advanced) robots creating the impression that they are agents in the sense of actually controlling and intending their own actions. Our results also suggest that –in case of NAO robots- failure, or even reluctance for doing tasks is received well, illustrating a promisingly positive view on robots.

Acknowledgements. We gratefully thank Luc Nies, Marc de Groot, Jan-Philip van Acken and Jesse Fenneman for their feedback and assistance in creating the videos for our study.

Appendix A: Description of the Videos

(1) **Giraffe.** Task description: a robot tidies up a room by picking up a toy giraffe, and dropping it in a toybox next to it.

LA. The robot tries picking up a toy giraffe, but as the robot lifts its body, the giraffe drops out of its hands. Despite this happening, the robot tries to complete the task by moving its arm to the toybox and opening its hands.

LE. The robot tries to picks up a toy giraffe, and looks at the toybox. Yet, instead of dropping the toy giraffe in the box, the robot throws it away from the box.

(2) **High Five.** Taskdescription: a robot asks a confederate how he is doing. When the confederate gives a positive reply, the robot says: "awesome, high five" and lifts its arm to give a high five.

LA. The robot has difficulty lifting its arm. As soon as the arm is up, it immediately drops to the ground; not even touching the arm of the confederate. After this happens, the robot states "oops".

LE. The robot lifts its arm. However, as soon at the confederate's arm is up, the robot tactically drops his arm by lowering it and pushing it underneath the confederate's arm. After this happens, the robot laughs.

(3) **Cardsorting.** Taskdescription: a robot is shown cards (one at a time) from a standard 52-deck of cards. Its task is to categorize the cards by naming the color on the card (hearts, diamonds, clubs, spades or joker).

LA. The robot starts off by naming a few cards correctly. However, after a while it starts stating some wrong colors (e.g. saying 'hearts' instead of 'spades'). Its timing is still correct.

LE. The robot starts off by naming a few cards correctly. However, after a while it ignores a card. After it has been quiet for a few seconds, the robot lifts its arms to shove the deck of cards away.

(4) Art. Task description: a robot sits in front of a table with a piece of paper on it. It holds a marker in its hand. Its task is to draw a house.

LA. The robot lowers the marker to draw something. Its arm makes movements as if it is drawing. However, due to giving too much pressure on the marker, the marker is restrained to the paper. Instead of a house, only a dot is drawn. The robot does not give notice of this problem.

LE. For a brief moment, the robot looks at the paper. However, instead of drawing, it lifts its face up again and throws the marker away.

(5) Dance. Task description: a robot states: "hello there! I'm a great dancer, would you like to see me dance?" After a confederate says: "yes", the robot continues: "alright! I will perform the Thai Chi Chuan Dance for you!". As follows, the robot starts to perform this dance.

LA. After stating that the robot will perform the dance, it starts playing a song (Chan Chan by Buena Vista Social Club). After a few seconds, the confederate states: "NAO you're not dancing". NAO immediately replies with: "oops! Wrong song. I will perform the Thai Chi Chuan Dance now." As follows, the robot starts to perform this dance.

LE. After stating that the robot will perform the dance, the robot starts playing a song (Chan Chan by Buena Vista Social Club). Meanwhile he states: "...or maybe not!" While sitting down, he concludes by saying "ha, let's take a break". The confederate tries to communicate with the robot by asking "Naomi? Why are you taking a break?", but it does not respond either verbally or non-verbally.

(6) Math. Task description: a robot solves some calculations out loud. For example, it says: "5 times 10 equals... 50!". This scenario does not include any further dialogue as introduction or conclusion.

LA. After a few calculations, the robot starts giving some wrong answers that imply that it mixes up the type of operation that is required. For example, it says: "120 divided by 3 equals... 123!".

LE. After a few calculations, the robot starts giving some useless (but possibly correct) answers, implying he does not feel like doing the task properly. For example, it says: "80 minus 20 equals...150! Divided by my age."

(7) Tictactoe. Task description: a robot plays a game of tic-tac-toe on a whiteboard with a human confederate. The robot is standing faced towards the whiteboard. The confederate is standing next to the whiteboard and will draw the X's and O's. As soon as the robot sees the confederate, it proposes to play the game. Accordingly, the game is successfully played and ends in a draw. The robot concludes by saying: "well, that was fun! Wanna play again?".

LA. After a few rounds, the robot asks the confederate to draw 'its' X on a block that is already filled with an X. The confederate corrects the robot and suggests it tries again. Consequently, the robot asks the exact same question. The confederate thickens the lines of the concerning X and asks the robot to try again. This time, the robot asks to fill an empty block, implying that, before, it did not see the X correctly. The game successfully continues and ends in a draw.

LE. After a few rounds, the robot asks the confederate to draw 'its' X on a block that is already filled with an O. The confederate corrects the robot and suggests he tries again. Consequently, the robot asks the exact same question. The confederate thickens the lines of the concerning O and asks the robot to try again. In response, the robot still repeats its question. When the confederate simply responds by saying: "No!", the robot responds with: "Ha, seems like you lost. But, practice makes perfect. Wanna play again?".

References

1. Van der Woerdt, S., Haselager, P.: When robots appear to have a mind: the human perception of machine agency and responsibility. New Ideas Psychol. (submitted)
2. Asaro, P.: Robot Ethics: The Ethical and Social Implications of Robotics. MIT Press, Cambridge (2013)
3. Singer, P.W.: Military robotics and ethics: a world of killer apps. Nature **477**(7365), 399–401 (2011)
4. Duffy, B.R.: Anthropomorphism and the social robot. Robot. Auton. Syst. **42**(3–4), 177–190 (2003)
5. Złotowski, J., Strasser, E., Bartneck, C.: Dimensions of anthropomorphism: from humanness to humanlikeness. In: Proceedings of the 2014 ACM/IEEE International Conference on Human-Robot Interaction (HRI 2014). ACM, New York (2014)
6. Schultz, J., Imamizu, H., Kawato, M., Frith, C.D.: Activation of the human superior temporal gyrus during observation of goal attribution by intentional objects. J. Cogn. Neurosci. **16**(10), 1695–1705 (2004)
7. Alicke, M.D.: Culpable control and the psychology of blame. Psychol. Bull. **126**(4), 556–574 (2000)
8. Epley, N., Waytz, A., Cacioppo, J.T.: On seeing human: a three-factor theory of anthropomorphism. Psychol. Rev. **114**(4), 864–886 (2007)
9. Weiner, B.: Judgments of Responsibility: A Foundation for a Theory of Social Conduct. Guilford Press, New York/London (1995)
10. Weiner, B., Perry, R.P., Magnussen, J., Kukla, A.: An attributional analysis of reactions to stigmas. J. Pers. Soc. Psychol. **55**(5), 738–748 (1988)
11. Epley, N., Waytz, A.: The Handbook of Social Psychology, 5th edn. Wiley, New York (2010)
12. Rudolph, U., Roesch, S.C., Greitemeyer, T., Weiner, B.: A meta-analytic review of help giving and aggression from an attributional perspective. Cogn. Emot. **18**(6), 815–848 (2004)
13. Coleman, K.W.: The Stanford Encyclopedia of Philosophy. Fall 2006 Edition. The Metaphysics Research Lab, Stanford (2004). http://stanford.library.sydney.edu.au/archives/fall2006/entries/computing-responsibility/
14. Vilaza, G.N., Haselager, W.F.F., Campos, A.M.C., Vuurpijl, L.: Using games to investigate sense of agency and attribution of responsibility. In: Proceedings of the 2014 SBGames (SBgames 2014), SBC, Porte Alegre (2014)
15. Moon, Y., Nass, C.: Are computers scapegoats? Attributions of responsibility in human computer interaction. Int. J. Hum. Comput. Interact. **49**(1), 79–94 (1998)

16. You, S., Nie, J., Suh, K., Sundar, S: When the robot criticizes you: self-serving bias in human-robot interaction. In: Proceedings of the 6th International Conference on Human Robot Interaction (HRI 2011). ACM, New York (2011)

17. Malle, B.F., Scheutz, M., Forlizzi, J., Voiklis, J.: Which robot am i thinking about? The impact of action and appearance on people's evaluations of a moral robot. In: Proceedings of the 11th International Conference on Human Robot Interaction (HRI 2016). ACM, New York (2016)

18. Malle, B.F., Scheutz, M., Arnold, T., Voiklis, J., Cusimano, C: Sacrifice one for the good of many? People apply different moral norms to human and robot agents. In: Proceedings of the 10th International Conference on Human Robot Interaction (HRI 2010). ACM, New York (2015)

19. Weiner, B., Kukla, A.: An attributional analysis of achievement motivation. J. Pers. Soc. Psychol. **15**(1), 1–20 (1970)

20. Weiner, B.: Intentions and Intentionality: Foundation of Social Cognition. MIT Press, Cambridge (1995)

21. Mantler, J., Schellenberg, E.G., Page, J.S.: Attributions for serious illness: are controllability, responsibility, and blame different constructs? Can. J. Behav. Sci. **35**(2), 142–152 (2003)

22. Caverley, D.: Android science and animal rights: does an anology exist? Connect. Sci. **18**(4), 403–417 (2006)

23. Schaerer, E., Kelly, R., Nicolescu, M.: Robots as animals: a framework for liability and responsibility in human-robot interactions. In: Proceedings of RO-MAN 2009: The 18th IEEE International Symposium on Robot and Human Interactive Communication. IEEE, Toyama (2009)

24. Gray, H.M., Gray, K., Wegner, D.M.: Dimensions of mind perception. Science **315**(5812), 619 (2007)

25. Heider, F.: The Psychology of Interpersonal Relations. Wiley, New York (1958)

26. Bakan, D.: The Duality of Human Existence: Isolation and Communion in Western Man. Rand McNally, Chicago (1956)

27. Trzebinski, J.: Action-oriented representations of implicit personality theories. J. Pers. Soc. Psychol. **48**(5), 1266–1278 (1985)

28. Weiner, B.: An attributional theory of emotion and motivation. Psychol. Rev. **92**(4), 548–573 (1986)

29. Mayer, R.C., Davis, J.H., Schoorman, F.D.: An integrative model of organizational trust. Acad. Manag. Rev. **20**(3), 709–734 (1995)

30. Jungermann, H., Pfister, H., Fischer, K.: Credibility, information preferences, and information interests. Risk Anal. **16**(2), 251–261 (1996)

31. Block, N.: Oxford Companion to the Mind, 2nd edn. Oxford University Press, New York (2004)

32. Graham, S., Hoehn, S.: Children's understanding of aggression and withdrawal as social stigmas: an attributional analysis. Child Dev. **66**(4), 1143–1161 (1995)

33. Greitemeyer, T., Rudolph, U.: Help giving and aggression from an attributional perspective: why and when we help or retaliate. J. Appl. Soc. Psychol. **33**(5), 1069–1087 (2003)

34. Waytz, A., Morewedge, C.K., Epley, N., Gao, J.H., Cacioppo, J.T.: Making sense by making sentient: effectance motivation increases anthropomorphism. J. Pers. Soc. Psychol. **99**(3), 410–435 (2010)

35. Peterson, C., Semmel, A., von Baeyer, C., Abramson, L.T., Metalsky, G.I., Seligman, M.E. P.: The Attributional Style Questionnaire. Cogn. Ther. Res. **6**(3), 287–300 (1982)

36. Avis, M., Forbes, S., Ferguson, S.: The brand personality of rocks: a critical evaluation of a brand personality scale. Mark. Theor. **14**(4), 451–475 (2014)

37. Friedman, B.: 'It's the computer's fault': reasoning about computers as moral agents. In: Proceedings of the Conference on Human Factors in Computing Systems. ACM, New York (1995)
38. Kahn Jr., P.H., Kanda, T., Ishiguro, H., Ruckert, J.H., Shen, S., Gary, H.R., Reichert, A.L., Freier, N.G., Severson, R.L.: Do people hold a humanoid robot morally accountable for the harm it causes? In: Proceedings of the 7th International Conference on Human Robot Interaction (HRI 2012). ACM, New York (2012)
39. Biswas, M., Murray, J.C.: Towards an imperfect robot for long-term companionship: case studies using cognitive biases. In: Proceedings of the IEEE/RSJ International Conference on Intelligent Robots and Systems (IROS). IEEE, New York (2015)

Performance Indicators for Online Secondary Education: A Case Study

Pepijn van Diepen and Bert Bredeweg[✉]

Informatics Institute, Faculty of Science,
University of Amsterdam, Amsterdam, The Netherlands
PepijnVanDiepen@gmail.com, B.Bredeweg@uva.nl

Abstract. There is little consensus about what variables extracted from learner data are the most reliable indicators of learning performance. The aim of this study is to determine such indicators by taking a wide range of variables into consideration concerning overall learning activity and content processing. A genetic algorithm is used for the selection process and variables are evaluated based on their predictive power in a classification task. Variables extracted from exercise activities turn out to be most informative. Exercises designed to train students in understanding and applying material are found to be especially informative.

Keywords: Learning analytics · Learning performance indicators

1 Introduction

Learning Analytics (LA) provides insight into the progress of students and their learning performance. It analyses learner data with the aim to improve the learning process. Whereas the potential of the field is promising, results are still preliminary. A common approach is to let the prediction of learning performance act as guidance for teachers to identify students that need intervention. Quantitative data concerning *resource use*, *time spent on resources* and *grades* have been used for the prediction of learning performance [7,14]. However, confidence about what data are most suited is limited [1,14].

The aim of this study was to determine what aspects of learning behaviour can be extracted from the log-data of a Learning Management System (LMS) in secondary education and are reliable indicators of learning performance. An extensive set of potentially valuable variables was composed and several rounds of selection were applied in order to find the most informative indicators.

2 Related Work

2.1 Relevant Variables

Several studies indicated the statistical relevance of *resource usage* as predictive variable, often in terms of usage counts [7,9]. The *time spent* on learning objects

© Springer International Publishing AG 2017
T. Bosse and B. Bredeweg (Eds.): BNAIC 2016, CCIS 765, pp. 169–177, 2017.
DOI: 10.1007/978-3-319-67468-1_12

(LOs) was also found to be an indicator of learning performance [8]. Variables concerning exercise behaviour such as the time spent on exercises, the number of successful and unsuccessful attempts, and scores were also reported to be related to learning performance [8,11,14]. Other studies found *study results* to be most informative [8,13], some reported *social interaction* being important [11], and numerous studies reported demographic data to be a reliable indicator [14,15].

The LMS used in our study offered a wide range of exercises and reading material. However, the inclusion of demographic data was prohibited due to privacy constraints and no data concerning social interaction was available.

2.2 Feature Selection Methods

When a large number of features is considered, a thorough feature selection process is essential to improve predictions, provide a deeper understanding of the case, guide the reduction of data, and yield simpler models [12]. A common initial means of feature ranking can be accomplished by analysing the Pearson correlation coefficients of features with the to be predicted variable. Univariate feature ranking can be preferable to multivariate feature selection methods due to its simplicity and scalability. However, features can be of more value when taken joined with other features. Univariate feature analysis does not detect such cases, Hence, multivariate feature selection methods should be considered [3]. Below the three main categories of feature selection algorithms are discussed.

Wrappers are simple yet robust feature subset selection methods that use prediction algorithm accuracies as a measure of subset quality. Wrappers search the entire space of feature subsets and therefore become computationally intractable when a large number of features is addressed. In order to cope with the scalability of wrapper methods, the search for feature subsets can be guided by search strategies such as Genetic Algorithms (GA). However, if wrapper methods are applied, the risk of overfitting increases.

Embedded methods are often faster solutions to feature selection since they embed the selection process into the training process and they use greedy search methods to address the problem of scalability. Greedy search strategies have the disadvantage that former decisions are never revisited, therefore they do not guarantee optimal solutions.

Filter methods are fast solutions to feature selection and are often used as a pre-processing step that uses general characteristics of the data to select features [12]. The advantage of filters is that the selection is made independent of the predictor that is used for the final prediction. Filter methods are often used for univariate feature analysis but in the context of multivariate feature analysis it is a reasonable approach to use a wrapper as a filter and train another often more complex predictor using the selected subset [3].

2.3 Prediction Methods

Numerous prediction algorithms have been used to classify learning performance on a *discrete* scale. Less research has been conducted concerning the prediction of learning performance on a *continuous* scale. Classification algorithms such as Decision Trees, Neural Networks, Naives Bayes, K-Nearest Neighbour, Support Vector Machines and Logistic Regression are regularly applied for LA [5,13]. Wolff et al. [15] used Decision Trees to predict whether a student would fail or pass a course. The accuracy of the predictive models varied from 0.77 to 0.98 over three different courses. This suggests that predictor performance could be course dependent. Macfadyen et al. [7] implemented a Binary Logistic Regression predictor in order to classify student failure. The classifier predicted student failure with an accuracy of 0.74. Minaei-Bidgoli et al. [8] used K-Nearest Neighbours and Decision Trees to classify student outcomes in terms of two and three learning performance classes. After the optimisation of algorithm parameters and by combining multiple classifiers an accuracy of 0.94 and 0.72 was achieved for the two- and three-classes respectively.

One could argue that simple prediction algorithms are preferred over more complex algorithms. This is because the decision making of simpler algorithms can be analysed better. These findings indicate that simple prediction algorithms such as Decision Trees and K-Nearest Neighbours can be successful in predicting learning performance.

3   Method

3.1   Data, Participants and Context

The data for the research presented in this paper was provided by educational publisher ThiemeMeulenhoff and was extracted from the logs of the online geography course De Geo[1]. De Geo is a geography course offering 1,166 exercises, 476 self-assessment tests and 9 chapters of reading material (i.e., the equivalent of a year of school material). The data consisted of chronological click logs (two months of data) and exercise results (7 months of data). The two datasets were combined to explore their full potential. Exercise data included the final score and a label stating whether the exercise was completed, incomplete or skipped.

The dataset included data of 226 first year, secondary education students from the Netherlands, aged 11–12. The course material included reading material (also referred to as *theory*), online exercises, and self-assessment tests. Each exercise was categorised according to Bloom's taxonomy for learning objectives [6]. Hence, 6 categories (*Remember* (89, 8%), *Understand* (139, 12%), *Apply* (676, 58%), *Analyse* (172, 15%), *Evaluate* (31, 3%) and *Create* (59, 5%)) which were hierarchically structured, meaning the mastery of the next category is supposed to follow from the mastery of the prior category. Exercise activity was analysed separately for each category.

Since all data was anonymous and no final grades were made available due to privacy constraints, learning performance had to be determined based on alternative sources. The self-assessment tests were designed to provide the students an

---

[1] https://www.thiememeulenhoff.nl/voortgezet-onderwijs/mens-en-maatschappij/aardrijkskunde/de-geo-onderbouw-9e-editie.

indication of their learning performance, therefore results on self-assessment tests were considered to be the most appropriate measure of learning performance. All students were labeled with the mean of their results on all self-assessment tests that they completed.

Due to the same privacy constraints the dataset is not publicly available.

3.2   Variable Selection

Composing an Initial Set. First, variables concerning overall online activity were considered (e.g., number of clicks, time online, theory/exercise time distribution). These variables were extracted from the data that was collected over all content together instead of specific types of content. Subsequently, content specific variables extracted from reading and exercise activities were considered. All data was categorised in terms of (i) exercise processing, (ii) theory processing, and (iii) overall behaviour. A set of variables was composed for each category based on the type of variables that were found to be reliable in the reviewed literature. In the case of the variables concerning exercise behaviour the data of each set of exercises belonging to a particular category was analysed individually. Additionally, all exercises were analysed when taken together as well. Two extraction methods were applied in order to address potential differences in difficulty between exercises. Method **A** assumes all exercises to be of equal difficulty and evenly time consuming, whereas method **B** does not. Method **B** compared and analyzed students' data per separate exercise while method **A** compared accumulated results per category.

Selection. Initially, a wide range of variables was included, followed by removing redundant and irrelevant variables from the set. A selection was made using a univariate variable selection method based on Pearson correlation with learning performance. All features that did not significantly correlate (p-value < 0.05) were discarded. Subsequently, multivariate variable selection was applied on the remaining variables. Embedded selection methods were not used since they rely on greedy selection algorithms which could exclude valuable features early in the process. Due to their computational complexity, wrappers based on the brute force methodology were also rejected. Therefore a combined filter/wrapper method as described in Sect. 2.2 was applied. A GA was used to guide the search for the best combination of variables[2]. GAs can be described as guided random search techniques that mimic the theory of evolution. They create populations of random individuals and select the best individuals to create the next population until an (sub)optimal solution is found. By using the prediction performance as fitness and variable subsets as individuals, GAs aim to select the strongest combination of variables.

For prediction a simple linear prediction model was selected as suggested by Guyon et al. [3]. The Linear Discriminant Analysis Classifier implemented by the Scikit Learn library [10] was selected for this purpose due to its simplicity and

---

[2] GA implementation from the DEAP library for evolutionary algorithms [2] was used.

low computational costs. The predictions were evaluated by a repeated 10-fold cross validation using 50 repetitions (see Sect. 3.3 for further explanation of these design choices). The algorithm was implemented following the guidelines provided by Fortin et al. [2]. The population consisted of 25 individuals, each representing a single feature subset. In each iteration of the evolutionary loop new offspring was generated by either mixing individuals using a uniform crossover method, mutating a single individual, or reproduction. As suggested by Fortin et al. the probability of mating individuals was set to 0.5 and the probability of mutating to 0.1. The mating of two individuals was accomplished using the uniform crossover method that exchanges the attributes of two individuals with an independent probability of 0.1. When an individual was mutated each feature was turned on or off with an independent probability of 0.05. Subsequently the offspring was joined with the original population and a selection made from the conjunction using the NSGA-2 selection operator provided by the DEAP library. The evolutionary loop was stopped at 75 iterations since that was the average point of convergence (stabilization of the population) from the test runs.

Since the feature subset space was searched extensively there was a significant probability that a combination of features was found that produces high predictive accuracy on the train set but would generalise poorly. Even when the fitness of individuals was determined by a cross-validation metric some overfitting could leak into the model. Since the dataset size was limited, and did not allow for a test set to be separated, 10-fold cross validation was applied to the GA feature selection process. The dataset was randomly split into ten parts, in a stratified fashion. After each iteration of the GA, the performance of the generated feature subset was tested on the validation set. The feature subset was optimised for the other 9/10 of the data and never saw the data in the validation set. Each fold had an optimal variable subset as output and a voting mechanism was used to make the final selection.

3.3 Classification

The predictive power of the selected variables was evaluated in two learning performance classification tasks: fail/pass and fail/sufficient/excellent. Classification algorithms such as Support Vector Machine (SVM), Gaussian Naive Bayes (GNB) and K-Nearest Neighbour (KNN) provided by the Scikit-Learn library [10] were used. All classifications were evaluated using repeated 10-fold cross validation. The k-fold Cross Validation (CV) estimator is a widely accepted model evaluation technique in the field of machine learning. Whereas it often produces unbiased estimates, the estimates can be highly variable when applied to a small dataset. Kim et al. [4] compared several bootstrap techniques to a repeated k-fold CV technique in order to address the problem of high variance in small datasets. They concluded that the repeated k-fold CV estimator outperformed bootstrap methods and recommended it for general use. Therefore the evaluation of all predictive models was conducted using a repeated k-fold CV, using $k = 10$ to maintain low bias. The number of repetitions was set to 50 because the model's confidence level stabilized at that point.

To find the optimal parameters for each learning algorithm the built-in grid search for optimal parameters of the scikit-learn library was used. It applies an exhaustive search trough a set of parameter options provided by the user to find the optimal parameters for the classifier.

All classifiers were evaluated in terms of accuracy, F2-score and recall of the class that represented the low performing students. Finally, a baseline classifier that was set to always predict the most common class was included in the evaluation.

4 Results

Over all variable categories together a total of almost 50 variables were considered. Within each category, variables concerning a student's time distribution, number of clicks and variations on those (such as ratios) were considered. Because variables extracted from different levels from Bloom's taxonomy were treated individually most variables originated from exercise behaviour. The selection process yielded a final set of 15 variables including almost all categories (Table 1). Two variables belonged to the overall activity category: total_clicks and the theory_exercise_ratio. Two variables belonged to the reading activity category: the theory_look_ups and the theory_look_ups_time. Eleven variables originated from the exercise activities, especially exercises from the *apply* category (five variables) and *understand* category (three variables) from Bloom's taxonomy were found to be reliable indicators. Notice that the variable exercise_incomplete_apply occurs twice in the list, once extracted using method **A** and once with method **B**. No variables came from the *remember* and *analyze* category. One variable came from both the *evaluate* and *create* category. From the exercise processing variables, the number of incomplete (wrong answer provided) was most informative, followed by the mean and total time spent on exercises. The last variable in the list concerned the mean time spent on an exercise over all of Bloom's categories together. To evaluate the predictive value of these variables, they were tested in two classification tasks. The baseline classifiers achieved an accuracy of 0.51 and 0.50 in the two and three class classification task respectively. The SVM classifier predicted most accurately for both classification tasks followed by GNB and KNN. An accuracy of 0.80 and recall of 0.84 of the *fail* class was achieved for the classification of two classes. For the classification of three classes an accuracy of 0.67 and recall of 0.67 was achieved. Other classification algorithms were also evaluated but resulted in less accurate predictions. However, all classifiers did perform significantly better ($p < 0.05$) than the baseline classifiers.

Table 1. Final selection of variables. Pearson correlation (r) with the learning outcome and corresponding p-values (p) are shown alongside with their ranking according to the GA (*votes*). Variable names end with A or B depending on the extraction method used.

#	Variable	Description	*votes*	r	p
1	exercise_incomplete_apply_A	Number of wrong answers on apply exercises	10	−0.52	0.00
2	theory_exercise_ratio	The ratio of time spent on exercises and theory	9	0.27	0.00
3	exercise_avg_time_understand_B	The mean time spent on understand exercises	9	0.23	0.00
4	exercise_time_evaluate_A	Total time spent on evaluate exercises	8	−0.16	0.04
5	exercise_completed_apply_A	Number of correct answers on apply exercises	8	0.16	0.04
6	theory_look_ups	Total number of theory look -ups during exercises	7	0.35	0.00
7	exercise_incomplete_understand_A	Number of wrong answers on understand exercises	7	−0.37	0.00
8	exercise_incomplete_apply_B	Number of wrong answers on apply exercises	7	−0.26	0.00
9	exercise_time_understand_A	Total time spent on understand exercises	6	0.23	0.00
10	total_clicks	Total number of clicks	6	0.17	0.03
11	exercise_avg_time_apply_A	The mean time spent on apply exercises	6	0.29	0.00
12	exercise_skipped_create_A	Number of skipped create exercises	6	−0.20	0.00
13	exercise_time_apply_A	Total time spent on apply exercises	5	0.24	0.00
14	exercise_time_A	Total time spent on exercises	5	0.20	0.00
15	theory_look_ups_time	Total time spent on theory look-ups	5	0.37	0.00

5   Conclusion

The aim of this study was to determine what LMS data best explains students' learning performance. In correspondence with the findings of Tempelaar et al. [14] the indicators concerning exercise processing were found to be most reliable. Variables extracted from exercise activities, that were designed to train students in *understanding* and *applying* material, were found to be especially informative. Both method **A** and **B** can be used to extract the variables although in general it seems that method **A** is sufficient. In contrast to Wolff et al. [15] variables describing general learning behaviour did contribute predictive value. Only theory processing variables related to exercise activity (look-ups during exercises) were part of the final variable list. This suggests that reading behaviour does not reveal much about learning outcome. However, a combination of features concerning overall activity, theory- and exercise-processing was needed to achieve the best prediction results. Therefore it is important to capture as many aspects of the learning process as possible in order to make accurate predictions.

To make predictions valuable for education, they need to be used to deliver valuable feedback to students. In our study none of the predictive models were analyzed in order to understand their decision making. In future research a better understanding of the predictions should be investigated and obtained.

Acknowledgements. The authors would like to thank ThiemeMeulenhoff for providing the resources for this study. Special thanks go to Joost Borsboom, Gilian Halewijn, Wouter van Rennes, Emiel Ubink and Johan Verhaar.

References

1. Esmeijer, J., van der Plas, A.: Learning Analytics en Zelfsturend Leren. TNO R10373 (2013)
2. Fortin, F.A., De Rainville, F.M., Gardner, M.A., Parizeau, M., Gagné, C.: DEAP: evolutionary algorithms made easy. J. Mach. Learn. Res. **13**, 2171–2175 (2012)
3. Guyon, I., Elisseeff, A.: An introduction to variable and feature selection. J. Mach. Learn. Res. **3**, 1157–1182 (2003)
4. Kim, J.: Estimating classification error rate: Repeated cross-validation, repeated hold-out and bootstrap. Comput. Stat. Data Anal. **53**, 3735–3745 (2009)
5. Kotsiantis, S., Pierrakeas, C., Pintelas, P.: Predicting students' performance in distance learning using machine learning techniques. Appl. Artif. Intell. **18**, 411–426 (2004)
6. Krathwohl, D.R.: A revision of Bloom's taxonomy: an overview. Theory Pract. **41**, 212–218 (2002)
7. Macfadyen, L.P., Dawson, S.: Mining LMS data to develop an early warning system for educators: a proof of concept. Comput. Educ. **54**, 588–599 (2010)
8. Minaei-Bidgoli, B.: Predicting student performance: an application of data mining methods with an educational web-based system. Comput. Educ. **47**, 157–167 (2015)
9. Morris, L.V., Finnegan, C., Wu, S.: Tracking student behavior, persistence, and achievement in online courses. Internet High. Educ. **8**, 221–231 (2005)
10. Pedregosa, F., Varoquaux, G., Gramfort, A., Michel, V., Thirion, B., Grisel, O., Blondel, M., Prettenhofer, P., Weiss, R., Dubourg, V., Vanderplas, J., Passos, A., Cournapeau, D., Brucher, M., Perrot, M., Duchesnay, E.: Scikit-learn: machine learning in python. J. Mach. Learn. Res. **12**, 2825–2830 (2011)
11. Romero, C., Ventura, S., García, E.: Data mining in course management systems: moodle case study and tutorial. Comput. Educ. **51**, 368–384 (2008)
12. Sánchez-Maroño, N., Alonso-Betanzos, A., Tombilla-Sanromán, M.: Filter methods for feature selection – a comparative study. In: Yin, H., Tino, P., Corchado, E., Byrne, W., Yao, X. (eds.) IDEAL 2007. LNCS, vol. 4881, pp. 178–187. Springer, Heidelberg (2007). doi:10.1007/978-3-540-77226-2_19
13. Shahiri, A.M., Husain, W.: A review on predicting student's performance using data mining techniques. Procedia Comput. Sci. **72**, 414–422 (2015)

14. Tempelaar, D.T., Rienties, B., Giesbers, B.: In search for the most informative data for feedback generation; Learning Analytics in a data-rich context. Comput. Human Behav. **47**, 157–167 (2015)
15. Wolff, A., Zdrahal, Z., Nikolov, A., Pantucek, M.: Improving retention: predicting at-risk students by analysing clicking behaviour in a virtual learning environment. In: Proceedings of the Third International Conference on LAK'33, pp. 145–149 (2013)

Demonstration Papers

SWISH DataLab: A Web Interface for Data Exploration and Analysis

Tessel Bogaard[1][✉] ⓘ, Jan Wielemaker[1,2] ⓘ, Laura Hollink[1] ⓘ,
and Jacco van Ossenbruggen[1,2] ⓘ

[1] Centrum Wiskunde & Informatica, Amsterdam, Netherlands
{Tessel.Bogaard,J.Wielemaker,L.Hollink,Jacco.van.Ossenbruggen}@cwi.nl
[2] Vrije Universiteit Amsterdam, Amsterdam, Netherlands

Abstract. SWISH DataLab is a single integrated collaborative environment for data processing, exploration and analysis combining Prolog and R. The web interface makes it possible to share the data, the code of all processing steps and the results among researchers; and a versioning system facilitates reproducibility of the research at any chosen point. Using search logs from the National Library of the Netherlands combined with the collection content metadata, we demonstrate how to use SWISH DataLab for all stages of data analysis, using Prolog predicates, graph visualizations, and R.

Keywords: Prolog · R · Data processing · Data mining

1   Introduction

Data is ubiquitous, and so are tools supporting data analysis. More often than not, different tools are used for different stages of the analysis. For example, the preprocessing and exploration of data is handled in one tool using one programming language and the analysis in a completely different environment, with scripts spread out in different files stored locally. This interferes with transparency, shareability and reproducibility of the research. SWISH DataLab[1] provides a web interface. It is a Wiki-like collaborative environment combining processing, exploration, and analysis of data, supporting transparency of the choices made. It blends the clarity of Prolog with the statistical computing power of R[2]. Data cleaning and creating concepts and abstractions over the data benefit from the elegance of Prolog's rule-based logic programming paradigm; the statistical analysis and visualization are the strength of R. The combination of logic programming and R has been shown in [1], and applied in the context of biomedical research, e.g., [2], and in sentiment analysis of social media, e.g., [3]. SWISH DataLab integrates SWI-Prolog and R into a single computational environment accessible through a shared web interface.

---

[1] A version of SWISH for teaching Prolog is available online: http://swish.swi-prolog. org/.

[2] https://www.r-project.org/.

© Springer International Publishing AG 2017
T. Bosse and B. Bredeweg (Eds.): BNAIC 2016, CCIS 765, pp. 181–187, 2017.
DOI: 10.1007/978-3-319-67468-1_13

Using SWISH DataLab, it is easy to quickly try out different data abstractions on a sample and evaluate the impact on the results. SWISH DataLab is currently being developed as a collaborative environment for responsible data science using data from the National Library of the Netherlands. In this case study we combine six months of search logs with the content metadata from the historical newspaper collection. We explore and process the datasets using the Prolog programming language, and use R for a statistical analysis of the data.

2   SWISH

SWISH DataLab is an instantiation of SWISH (SWI-Prolog for SHaring), geared to data analysis. In SWISH, a web interface gives access to the computational environment. The development of SWISH [4,5] is influenced by Jupyter[3] and JSFiddle.[4] With JSFiddle it shares the model of a server where people can save and share programs and documents. From Jupyter the concept of *notebooks* is taken, a mixture of text and program fragments that can be edited in a browser [6].

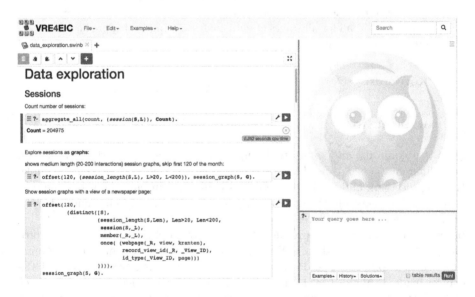

Fig. 1. The interface of SWISH DataLab, showing a Prolog notebook with some queries on the left and a query window on the right.

In SWISH, programs can be executed from the browser and the results appear in the browser as plain answers or rendered as tables, graphs or charts (Fig. 1). The core language of SWISH is SWI-Prolog. Unlike Jupyter it does not support

---

[3] http://jupyter.org/.
[4] https://jsfiddle.net/.

other languages directly. Instead, it allows for extending the Prolog core by binding it to other systems. On the *backend* it may be connected to external data using e.g., ODBC or SPARQL. Computational flexibility can be extended using e.g., R. The complementary power resulting from combining logic programming and R has been demonstrated in [1]. The *frontend* can be enhanced using server-side support from e.g., R or Graphviz[5] as well as client-side support using e.g., D3.js[6] or C3.js[7] as visualization methods.

2.1   SWISH and the R Programming Language

R is made available to Prolog by means of the Rserve[8] package. This implies that every SWISH user has a private instance of R, providing both isolation and concurrency.

R can be accessed from Prolog using two distinct mechanisms. The predicate *Result <- Expression* uses Prolog syntax to represent R expressions (Fig. 2). The match is close, but not 100%. For example, where '10.' is a valid R floating point number, Prolog requires writing this as '10.0' and R identifiers that start with a

```
1  :- include(session_viz).
2  :- <- library("ggplot2").
3  :- <- library("scales").
4
5  ## create_barchart
6  %  exploration of data
7  create_barchart :-
8      x_label(Name),
9      title(Title),
10     r_data_frame(data, [value=Value, count=Count],
11              aggregate(count, limit(20000, data_count_value(Value)), Count)),
12     data$percentage <- round(data$count/sum(data$count) * 100, digits=2),
13     barchart(data, value, Name, Title).
14
15 barchart(Data, Column, XLab, Title) :-
16     <- ggplot(Data, aes(x=factor(Column), y=count, fill=Column))
17     ) + geom_bar(stat = "identity", size = 0.2
18     ) + xlab(XLab) + ylab("Frequency")
19     ) + ggtitle(Title
20     ) + scale_y_continuous( labels = comma
21     ) + theme(plot.title = element_text(size=16),
22              axis.title=element_text(size=14),
23              axis.text.x = element_text(size = 10, angle = 30, hjust = 1),
24              axis.text.y = element_text(size = 10)
25     ) + guides(fill=false
26     ) + geom_text(colour = "black", size = 5, aes(label=percentage), vjust = -0.3).
27
28 ## create histogram
```

R code in Prolog program

Fig. 2. R expression in a Prolog background program in SWISH DataLab.

---

[5] http://www.graphviz.org/.

[6] https://d3js.org/.

[7] http://c3js.org/.

[8] https://www.rforge.net/Rserve/.

capital letter need single quotes to avoid misinterpretation as a Prolog variable. Complex R objects such as functions cannot be expressed using the Prolog syntax. This problem is resolved using *quasi quotations* [7]. Quasi quotations allow for embedding external languages verbatim, while interpolating values from Prolog. The basic syntax is {|lang(param...)||code|}. For example, we can write ? − {|r||plot(c(1, 2, 3))|}. to realize a simple R plot. We can combine this with Prolog as shown:

```
?- numlist(1, 25, Data),
   {|r(Data)||plot(Data)|}.
```

Quasi quotations allow for reusing long snippets of R code verbatim, while the Prolog syntax is more natural for relatively simple R calls and allows for building R calls dynamically.

3   Case Study: Analysis of Online User Search Behavior

The goal of this use case is to understand user search behavior. Under strictest confidentiality agreement we have received six months of server logs from the full text search platform[9], spanning a period from October 2015 to March 2016. This search platform provides access to combined collections from the National Library of the Netherlands and other national heritage and research institutes. These collections are–as is the case for other digital libraries and archives–characterized by bibliographic data describing the content (e.g. publication date, type of document, origin of document). These metadata values are reflected in the search interface in *facets* that can help filter the results (Fig. 3). Over 90% of user requests accesses the historical newspaper collection. For this reason we focus on this collection, that contains over 100M documents across four centuries. Using SWISH DataLab, we have linked the content metadata values to the clicked and downloaded documents, enabling a comparison between facet-use in search and the metadata of clicks and downloads.

3.1   Iterative Data Exploration Through Abstraction

Constructing a vocabulary. The goal to describe user search behavior has motivated an exploration of the server logs based on user interactions within sessions. The sessions are defined based on IP address and a 30 min timeout. In order to recognize usage patterns, we have visualized these sessions in graphs with Graphviz for the rendering (see Fig. 4). As an added benefit, the graphs help to conceptualize the data.

Visualization of abstractions. The graphs make visible that users visit the same search engine results page often in a session, leading to a Prolog rule where we abstract this to a single node with multiple incoming arrows. This abstraction

---

[9] http://www.delpher.nl/.

Fig. 3. Search interface of the Delpher platform, with facets on the left and results to the right.

produces a different count of how many search interactions a user has engaged in, a count where revisiting the same results page in the same session is not seen as a new search (Fig. 5).

Removing reloads. We also removed repeated visits of the same web page right after each other, as this is likely a reload of the web browser and not a new interaction by the user. This has resulted in a more practical count of the number of clicks and a clearer definition of the dwell time on a document (as we time this from the first load of the page until a new interaction and not as separate shorter dwell times).

Return to any previous state. These incrementally written rules to clean the data and visualize the graphs can be fully traced in the web environment, where all intermediate saved versions of both data and code have been saved and can be retrieved. Being able to return to an exact previous state improves the reproducibility of the research, making it possible to save the state of the project at the time of publication and to rerun the exact steps on the same (or possibly updated) data.

Evaluate concepts and abstractions. We can inspect the statistical effect of different symbolic definitions of the data, such as shown in Fig. 5, where two abstractions over the data are set aside each other in a single notebook. The environment supports this type of transparency in code and results, making it possible to measure the impact of these abstractions on the results.

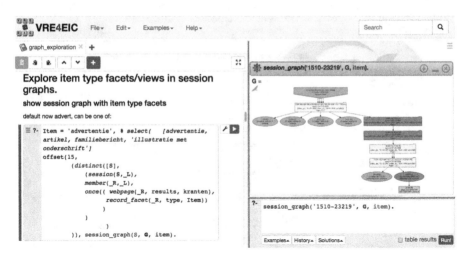

Fig. 4. Visualizing user sessions in graphs. Query code on the left, session graph to the right.

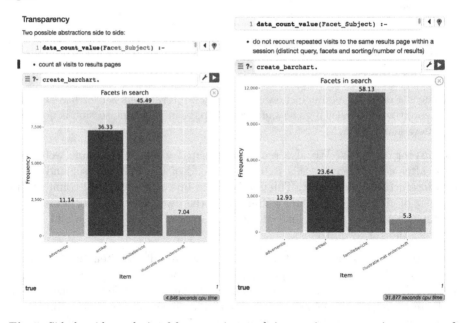

Fig. 5. Side by side analysis of facet use in search interactions, comparing a count of all search interactions versus not recounting revisits within a session.

4   Conclusions and Future Work

With SWISH DataLab we are moving closer to a single environment for responsible data science shared between researchers. The use of visualizations in combination with concepts and abstractions defined as rules makes data cleaning more transparent and more thorough, and patterns in user interactions more insightful.

Future work on the use case will include a better definition of a session based on the graphs and machine learning (using for example the machine learning algorithms available in R) for prediction of behavior.

SWISH DataLab will be extended with facilities to improve collaboration such as sending change notifications, shared editing and a commenting/chat service. We also plan to facilitate generating a permanent link that captures a result (e.g., a chart or table) and all programs and data needed to reproduce this result reliably.

Acknowledgments. We thank the National Library of the Netherlands for their support. The development of SWISH DataLab was partially supported by the VRE4EIC project, a project that project has received funding from the European Union's Horizon 2020 research and innovation programme under grant agreement No 676247.

References

1. Angelopoulos, N., Santos Costa, V., Azevedo, J., Wielemaker, J., Camacho, R., Wessels, L.: Integrative functional statistics in logic programming. In: Sagonas, K. (ed.) PADL 2013. LNCS, vol. 7752, pp. 190–205. Springer, Heidelberg (2013). doi:10.1007/978-3-642-45284-0_13
2. MacIntyre, D.A., Chandiramani, M., Lee, Y.S., Kindinger, L., Smith, A., Angelopoulos, N., Lehne, B., Arulkumaran, S., Brown, R., Teoh, T.G., Holmes, E., Nicoholson, J.K., Marchesi, J.R., Bennett, P.R.: The vaginal microbiome during pregnancy and the postpartum period in a European population. Sci. Rep. **5**, 8988 (2015). EP
3. Andreasen, T., Christiansen, H., Have, C.T.: Querying sentiment development over time. In: Larsen, H.L., Martin-Bautista, M.J., Vila, M.A., Andreasen, T., Christiansen, H. (eds.) FQAS 2013. LNCS (LNAI), vol. 8132, pp. 613–624. Springer, Heidelberg (2013). doi:10.1007/978-3-642-40769-7_53
4. Wielemaker, J., Lager, T., Riguzzi, F.: SWISH: SWI-Prolog for sharing. CoRR abs/1511.00915 (2015)
5. Beek, W., Wielemaker, J.: SWISH: an integrated semantic web notebook. In: Kawamura, T., Paulheim, H., (eds.) Proceedings of the ISWC 2016 Posters & Demonstrations Track co-located with 15th International Semantic Web Conference (ISWC 2016). CEUR Workshop Proceedings, Kobe, Japan, 19 October 2016, vol. 1690. CEUR-WS.org (2016)
6. Ragan-Kelley, M., Perez, F., Granger, B., Kluyver, T., Ivanov, P., Frederic, J., Bussonier, M.: The Jupyter/iPython architecture: a unied view of computational research, from interactive exploration to communication and publication. In: AGU Fall Meeting Abstracts, vol. 1, p. 07 (2014)
7. Wielemaker, J., Hendricks, M.: Why it's nice to be quoted: Quasiquoting for Prolog. CoRR abs/1308.3941 (2013)

Author Index

Printed in the United States
By Bookmasters